Springer Series in Optical Sciences Volume 47

Edited by Arthur L. Schawlow

Springer Series in Optical Sciences

Volume 42 **Principles of Phase Conjugation**
By B. Ya. Zel'dovich, N. F. Pilipetsky, and V. V. Shkunov

Volume 43 **X-Ray Microscopy**
Editors: G. Schmahl and D. Rudolph

Volume 44 **Introduction to Laser Physics**
By K. Shimoda

Volume 45 **Scanning Electron Microscopy**
By L. Reimer

Volume 46 **Holography in Deformation Analysis**
By W. Schumann, J.-P. Zürcher, and D. Cuche

Volume 47 **Tunable Solid State Lasers**
Editors: P. Hammerling, A. B. Budgor, and A. Pinto

Volumes 1-41 are listed on the back inside cover

Tunable Solid State Lasers

Proceedings of the First International Conference
La Jolla, Calif., June 13–15, 1984

Editors:
P. Hammerling, A. B. Budgor, and A. Pinto

With 147 Figures

Springer-Verlag
Berlin Heidelberg New York Tokyo 1985

PETER HAMMERLING, Ph. D.
La Jolla Institute, P.O. Box 1434, La Jolla, CA 92038, USA

AARON B. BUDGOR, Ph. D.
Northrop Electronics Division, 2301 West 120th Street, Hawthorne, CA 90250, USA

ALBERT PINTO, Ph. D.
US Army Night Vision & Electro Optics Laboratory, Fort Belvoir, VA 22060, USA

ISBN 3-540-15135-4 Springer-Verlag Berlin Heidelberg New York Tokyo
ISBN 0-387-15135-4 Springer-Verlag New York Heidelberg Berlin Tokyo

Offset printing: Beltz Offsetdruck, 6944 Hemsbach/Bergstr. Bookbinding: J. Schäffer OHG, 6718 Grünstadt.
2153/3130-543210

Preface

In recent years there has been renewed interest in the scientific and industrial communities in tunable solid state vibronic lasers. Much of this has been spurred by the user desirous of obtaining compact primary laser sources (independent of nonlinear optical frequency shifts) throughout the visible and near infra-red spectral regions. To further motivate and stimulate research and development in this area, workshops sponsored by the Laser Division of the US Army Night Vision Electro-Optics Laboratory (NVEOL) at Fort Belvoir, Virginia were held during 1-3 April 1981 and 16-17 June 1983 at Keystone and NVEOL, respectively. The consensus of opinion of the participants at these workshops was that any successful program leading to the identification and development of vibronic tunable laser materials in the visible and IR must include coordinated activity between crystal growers, theoretical and experimental investigators into the fundamental processes of vibronic lasing, and laser device engineers. Continued interaction between government, industry, and academia was encouraged in order to establish a unified approach to these areas and, when necessary, redefine and redirect programmatic activity.

The organization of this 1st Annual Conference on Tunable Solid State Lasers held at the La Jolla Institute 13-15 June 1984, was based around the latest results in tunable vibronic materials and laser development, but structured in a manner consistent with the recommendations of the NVEOL workshops. The emphasis of the meeting focused on the major issues surrounding the discovery and development of solid state visible and IR vibronic lasers, and attempted to discern unified themes in the generic areas of:

1. Laser host/dopant identification and growth procedures,
2. Theoretical analysis to elucidate fundamental tunable laser principles,
3. Experimental investigations on laser spectroscopy to which theoretical analyses and models can be anchored,
4. Auxiliary technology development on efficient laser pumping sources (diodes, flashlamps), and
5. Reliability and standards definitions.

Specific attention was given to those items of greatest concern to tunable laser development, viz., theoretical understanding of the complex interactions between electrons, phonons, and crystal structure and the mechanisms leading to the prediction of solid state tunability; and rapid growth of crystal media and doping to accelerate the demonstration of new candidate tunable lasers. To enhance theoretical understanding of complex optical vibronic solids, an attempt was made to induce cross-fertilization of ideas and tools used in describing optical phenomena in amorphous materials and in semiconductors. Invitees at the theoretical forefronts of these fields presented tutorials which will hopefully spur new development in optical properties of vibronic lasers.

The workshop structure was chosen for this conference in order to promote an informal atmosphere and thus maximize attendee interaction. Twenty invited presentations, spread over 2-1/2 days in five sessions, were given by experts in the areas of solid state crystal growth and lasing ion doping techniques, theoretical and experimental topics relating to vibronic tunability, and interdisciplinary concepts. A total of seventeen contributed poster papers were also incorporated.

The conference editors would like to take this opportunity to thank the sponsors, R. Guenther (Army Research Office), R. Buser (Army Night Vision Electro-Optics Laboratory), and F. Allario (NASA Langley Research Center); Session Chairpersons, L. Marquet (DARPA), R. Buser (NVEOL), C.R. Jones (ARO), W. Krupke (Lawrence Livermore National Laboratory), R. Orbach (UCLA), and J. Paul (NVEOL); Technical Advisory Committee, M. Weber (DOE), R. Powell (Oklahoma State), A. Budgor (Northrop Corporation), P. Moulton (Lincoln Laboratories), A. Linz (MIT), and N. Winter (Lawrence Livermore National Laboratory): and Round Table participants for their organized and outstanding efforts in making this meeting the success it undoubtedly was. Special thanks go to the administrative efforts of G. Pitts of the La Jolla Insitute, and V. McDuffee of Allied Technologies.

December 1984 *P. Hammerling · A.B. Budgor · A. Pinto*

Contents

Introduction

The conference on tunable solid state lasers discussed, in particular, operational vibronic lasers, advanced designs and applications, vibronic materials spectroscopy, crystal growth, vibronic laser theory, and interdisciplinary topics.

At the conclusion of the conference, a Round Table panel comprised of F. Allario (NASA Langley), R. Byer (Stanford), H. Jenssen (MIT), D. McClure (Princeton), R. Orbach (UCLA), G. Loiacono (Phillips), P. Moulton (Lincoln Laboratories), and M. Weber (DOE), reviewed the information presented and recommended future directions for focused research and development encompassing the five generic areas listed above. Their discussion is summarized below; for further details, the reader is referred to the Section indicated.

1. Ni-, Co-, V-Doped Hosts (Moulton) (Section I)

The V-, Co-, and Ni-doped fluorides and oxides fluoresce at greater than 1 μm -- V:$KMgF_3$ (λ_p = 1.05μm); Ni:MgO/MgF_2 (λ_p ~ 1.3/1.6μm); Co:MgF_2 (λ_p ~ 1.9μm). None of these laser crystals, however, have demonstrated good performance at room temperature. Both V- and Ni-doped materials exhibit large ESA and fluorescence quenching by non-radiative multi-phonon emission. Theoretical prediction of ESA has not been successful. Although Co:MgF_2 is a fairly good laser at low temperature (80 K), fluorescence quenching plays a dominant role for poor laser performance at temperatures around 225 K. ESA does not appear to be a particularly deleterious factor.

2. Crystal Growth (Jenssen, Loiacono, Weber) (Section III)

To date, the Czochralski method has been used most extensively to grow YAG, Al_2O_3, $KMgF_3$, GSGG, and alexandrite of sufficient dimensions for incorporation into laser devices utilizing rods and slabs. Other techniques for congruently melting materials are available – such as Bridgman and Heat Exchanger Method (HEM) – the former excellent for preparing spectroscopic study samples of metal halides and the latter a variant of Bridgman which, although it has the potential for large-volume, low-cost production of oxides and fluorides, is still in engineering development. Since the maturation of crystal growth techniques and subsequent improvements in laser crystal quality requires at least a decade, it would be advantageous to achieve collaborative interaction between theoreticians, crystal growers, and laser researchers and engineers to jointly guide and influence the directions of tunable laser R&D. Theoretical insight into what combinations of ion/host materials would exhibit tunability over pre-selected wavelength ranges with specified performance levels could eliminate much of the guesswork from laser invention. Rapid growth "intrinsic defect" – scattering centers, secondary precipitates, and impurities – free samples could enable the user community to preview the physical properties of the material, as well as such laser characteristics as fluorescence life-time, gain, and excited state absorption. Guidance could then be provided, with feedback, as to actual requirements.

3. Theoretical Aspects (McClure, Orbach) (Section IV)

A proper theory of vibronic tunable lasing needs to: take into account the actual offset of the energy states within the configuration coordinate space in order to properly describe excited state absorption (ESA); utilize an atomic cluster basis representation including the effects of lattice distortions to accurately depict charge transfer processes; incorporate a sophisticated model of electron-phonon interactions to elicit the fundamentals of absorption and fluorescence; and include, in addition to the totally symmetric mode, a number of asymmetric vibrational modes coupling to electronic states in order to predict the temperature dependence of fluorescence and non-radiative relaxation times. Though these theoretical developments are fairly major undertakings, success in predictive capability will significantly aid both crystal grower and laser experimenters in their activities.

4. Cross Disciplinary Topics (Allario, Byer, Loiacono, Weber) (Section IV)

A major problem facing the crystal growth community is the lack of students and advocate support from government and industry. A possible mechanism suggested to revitalize this discipline, as well as aid the crystal growth/laser spectroscopy and device engineering/theoretical troika in communicating with each other and, thus, being more responsive to each other's needs is to establish a center of excellence at a university or government laboratory which will be responsible for lead R&D coordination in these areas. This should enable a critical mass of researchers to be assembled which can then be tapped as a national resource.

A consensus of opinion was reached by the panel that device engineering issues have not been adequately addressed at this meeting. Better pump sources, improvements in crystal optical quality, reliability, and longevity are topics which are ever present when constructing a laser. Flashlamp efficiencies are on the order of 10 percent; this can be increased to 35 percent for access to red pump bands by employing GaAs diodes. For many applications, optical quality needs to be improved. Nd:YAG optical quality is consistently $\lambda/4$ per inch. This is to be compared with the optical quality of fine glass — $\lambda/10$ over 30 cm.

With regard to the eventuality of tunable vibronic lasing in glass host materials, the inhomogeneously broadened linewidth and poor thermal conductivity seem to bias against attainment of repetition rated, high quantum efficient devices.

Part I

Operational Vibronic Lasers
Advanced Design and Applications

Recent Advances in Transition-Metal-Doped Lasers

P.F. Moulton

Lincoln Laboratory, Massachusetts Institute of Technology
Lexington, MA 02173-0073, USA

Two broadly tunable, transition-metal-doped solid state lasers will be discussed in this paper. One uses crystals of $Co:MgF_2$ and operates between 1.5 and 2.3 µm; the other employs $Ti:Al_2O_3$ to obtain output from 0.66 to 0.99 µm. The $Co:MgF_2$ laser was first demonstrated in the early years of laser research [1,2], and our studies of this system have been directed towards a better understanding of the basic laser parameters as well as development of a device suitable for a variety of tunable-laser applications. The $Ti:Al_2O_3$ device is one of the newest solid state tunable lasers and is the first of perhaps a class of systems based on the Ti^{3+} ion.

1. $Co:MgF_2$

Our efforts on $Co:MgF_2$ lasers have involved operation with cryogenically cooled crystals optically pumped by cw or pulsed Nd lasers operating around 1.3 µm. Typical experiments were carried out using 6 x 6 x ~ 35 mm Brewster-angle crystal doped with ~ 1 wt.% CoF_2. The laser configuration was a longitudinally pumped two-mirror cavity, with only the TEM_{00} mode of the $Co:MgF_2$ laser excited by the pump. The input-output energy curves under pulsed pumping conditions for three different sets of cavity mirrors with coatings optimized for differing wavelengths are shown in Fig. 1. Data was taken at a pulse repetition rate of 4 Hz; the output pulsewidth well above threshold was ~ 1 ms. Threshold pump input energy at the 1.92-µm operating wavelength was 4.5 mJ. Measurements of the transmission losses of the pump focusing lens and cavity mirror and of the transmission of the laser crystal showed that approximately 46% of the pump input energy was absorbed in the crystal. When this is taken into account, the calculated output quantum efficiency (i.e., the ratio of laser output photons to absorbed pump photons) is 79% at the 1.92-µm operating wavelength. It is worth noting that this efficiency was obtained with a 0.9%-transmission output coupler, an indication that the scattering losses in the laser crystal were extremely small. The pulsed laser tuning curves for four different sets of cavity mirrors are shown in Fig. 2.

Pulsed operation from the $Co:MgF_2$ laser is possible at temperatures higher than that of liquid nitrogen, but the pump pulsewidth must be reduced to be consistent with the shorter upper-state lifetime of the Co^{2+} ion. The input-output curves for a crystal cooled to 225 K on a two-stage thermoelectric cooler are shown in Fig. 3, compared to results under similar pumping conditions but with the crystal cooled to 80 K. Such high-temperature operation opens up a number of possible applications for the $Co:MgF_2$ laser which may have been precluded because of the former need for liquid nitrogen.

This work was sponsored by the Department of the Air Force.

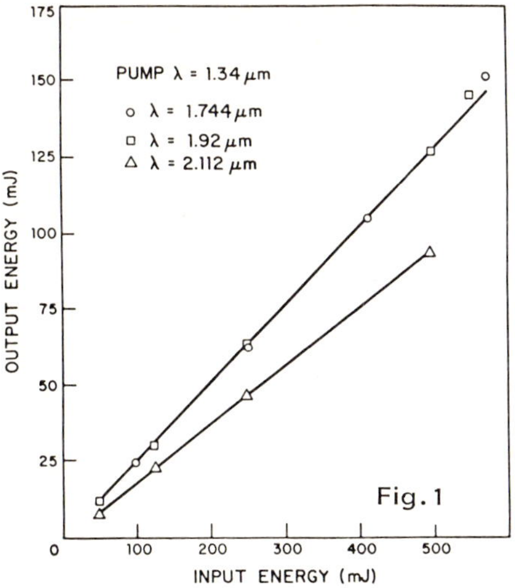

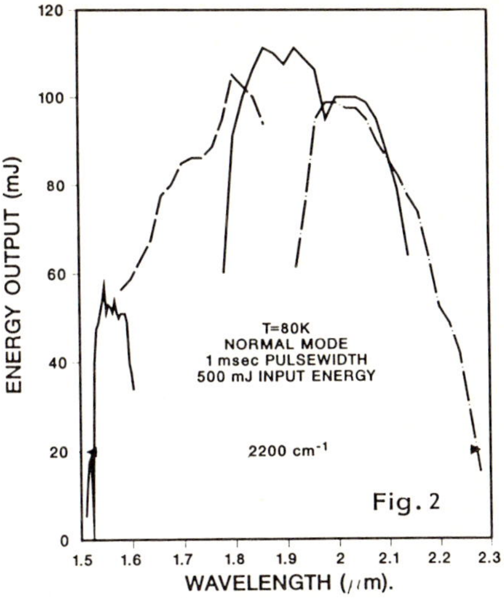

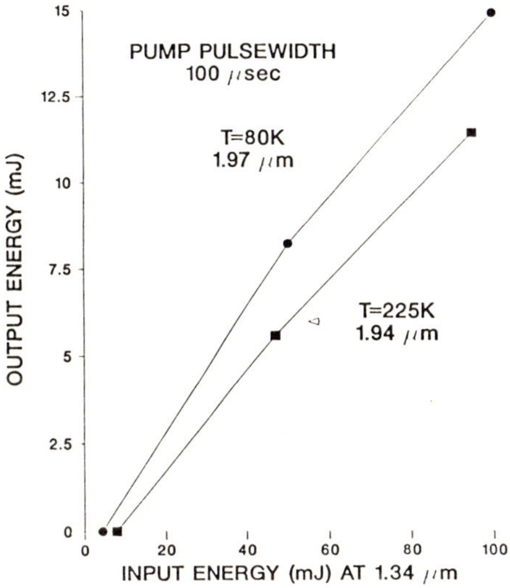

Fig. 1. Input-output energy curves for a pulsed $Co:MgF_2$ laser at three different output wavelengths. Output-mirror transmissions were 0.9% for the three wavelengths shown

Fig. 2. Composite tuning curve for pulsed $Co:MgF_2$ laser, based on results with four different sets of cavity mirrors. Output-mirror transmissions were 0.9% except for the shortest-wavelength curve, obtained with a 0.5%-transmission mirror

◂ Fig. 3. Input-output energy curves for pulsed $Co:MgF_2$ laser at two crystal temperatures. The fluorescence lifetime of $Co:MgF_2$ changed from 1.3 ms at 80 K to 0.3 ms at 225 K

Theoretical analysis of the normal-mode performance at ~ 80 K crystal temperature shows the peak laser gain cross section to be 1.5×10^{-21} cm^2, with crystal losses determined to be in the range of $1\text{-}5 \times 10^{-4}$ cm^{-1}. Excellent agreement between the observed and predicted temporal behavior has been found, indicating the $Co:MgF_2$ laser dynamics are well understood. Figure 4 compares the observed and calculated output waveforms, with the pumping ratio over threshold being the only adjustable parameter in the theoretical calculation. The theoretical pumping ratio providing the best match with the experimental data was 3.15, in good agreement with the experimental ratio of approximately 3.

A pulse-pumped, Q-switched Co:MgF_2 laser has generated TEM_{00}-mode energies of up to 60 mJ in a 150-ns-long pulse. Q-switched output was limited by optical damage to either the laser crystal or the cavity mirror surfaces. The Q-switched output energy was greater than the normal-mode energy for the same pump energy at low pumping ratios over threshold, but dropped below the normal-mode energy at pump ratios greater than ~ 2. Figure 5 shows the input-output energy curves for normal-mode and Q-switched operation with two output-mirror transmissions. Analysis of the Q-switched data indicates that intracavity peak intensity and not fluence was important in setting the optics damage level, and mirror damage appeared at peak intensity levels of 350 MW/cm^2. Reasonably good agreement was found between theory and experiment in predicting pulse buildup time and pulsewidth, with theoretical predictions of the latter tending to yield shorter pulses than observed. Q-switch output energy as a function of Q-switch on-time was found to occur at a somewhat earlier time and fall off faster than predicted. Theoretical calculations confirmed that Q-switched energy should be greater than normal-mode energy at low pump ratios, but no conclusive explanation was found for the rolloff in Q-switched energy at high pump ratios.

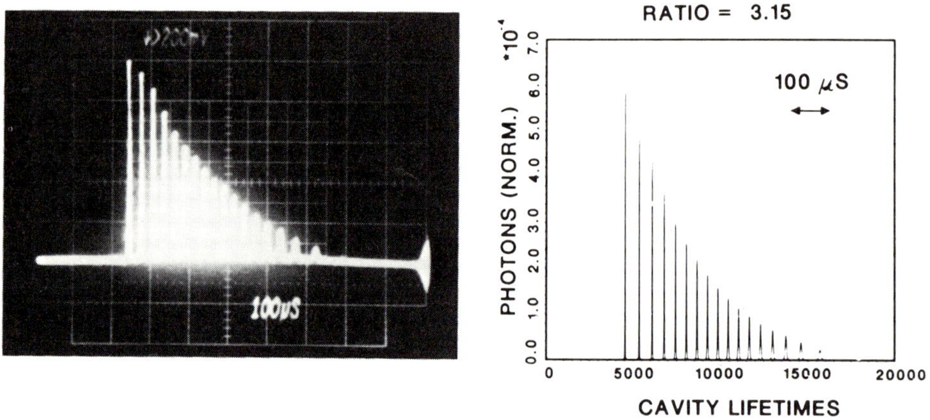

Fig. 4. Comparison of observed pulse train from normal-mode Co:MgF_2 laser with theoretical prediction

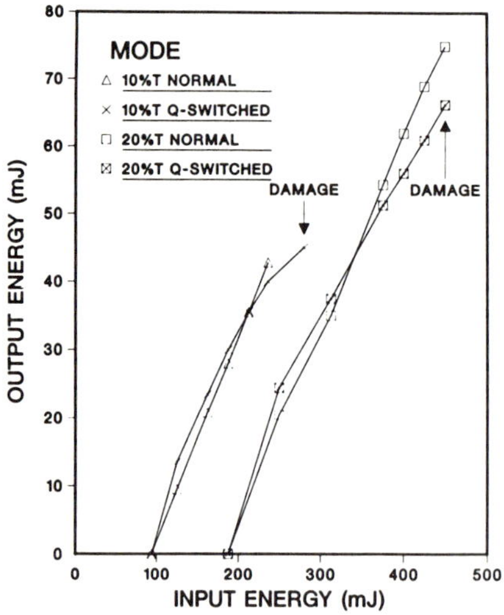

Fig. 5. Input-output energy curves for Co:MgF_2 laser in both normal and Q-switched modes, for two different output mirror transmissions

2. $Ti:Al_2O_3$

The degenerate energy levels of a single d-electron of isolated Ti^{3+} are split to first order in a crystalline environment into a $^2T_2$ triply degenerate lower level and a doubly degenerate ^2E upper level. Because of the strong phonon coupling, 4-level laser laser operation using the two d-electron levels is possible but excited-state absorption to higher d-electron levels is eliminated. The possibility of ESA to higher non-d-electron states, such as p-states or charge-transfer states still exists, but usually such levels are too high in energy to have any effect. The measured d-electron, π-polarized absorption and fluorescence bands from $Ti:Al_2O_3$ at 300 K are shown in Fig. 6. Structure in the absorption band is due to Jahn-Teller splitting of the ^2E upper level. The doping level of the first sample studied was measured by x-ray fluorescence techniques to be approximately 0.1 wt% Ti_2O_3; from this the peak absorption cross section was determined to be at least 3.2×10^{-20} cm^{-2}. The fluorescence lifetime was found to be 3.2 μs at 300 K, increasing to 3.8 μs at 77 K.

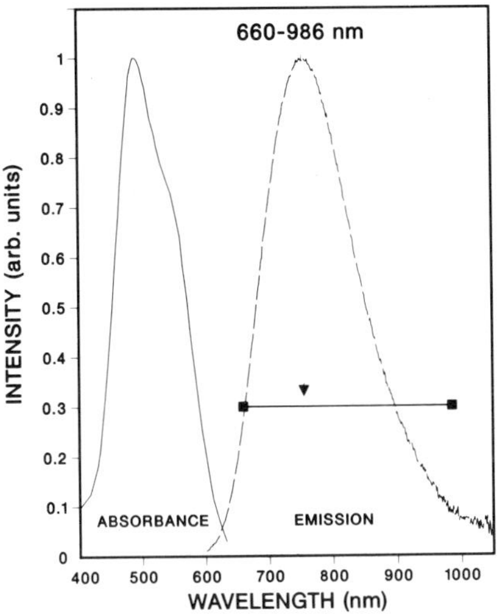

Fig. 6. Absorbance and emission spectra from $Ti:Al_2O_3$ at room temperature. The tuning range of laser is also indicated

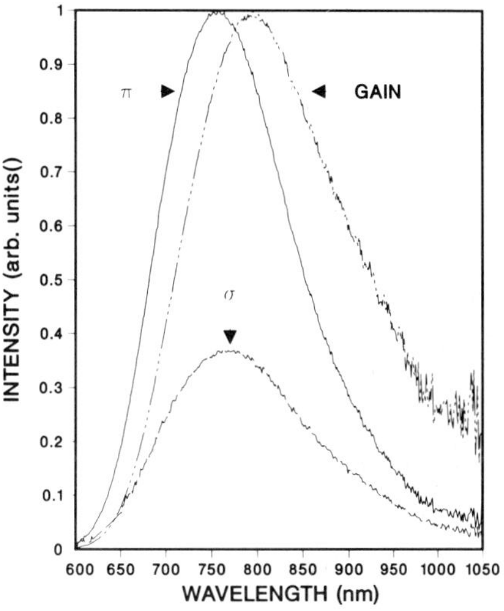

Fig. 7. Polarized emission spectra from $Ti:Al_2O_3$, along with theoretical gain lineshape

The laser tuning range, shown superimposed on the emission band in Fig. 6, extends out to longer wavelengths than one might expect from the emission lineshape. An analysis of the Einstein relation connecting spontaneous emission and gain cross section shows that if the emission spectrum is measured in watts per unit wavelength interval (as is the case in this discussion), the proper lineshape for the gain cross section is derived by multiplying the emission lineshape by the wavelength to the fifth power. The gain lineshape calculated in this manner is shown in Fig. 7, which also compares the relative emission intensities for the π and σ polarizations. The wavelength correction applied to the emission spectrum causes a significant increase in predicted gain at long wavelengths.

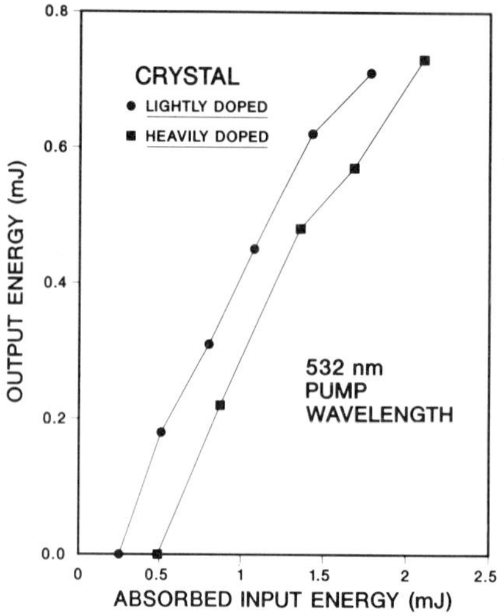

Fig. 8. Input-output energy curves for $Ti:Al_2O_3$ laser pumped by frequency-doubled Nd:YAG laser

Laser operation from $Ti:Al_2O_3$ was first obtained by longitudinally pumping a crystal with the output of a ~ 500-nm-wavelength, coaxial-flashlamp-pumped dye laser. Output energies at 750 nm of 7 mJ, at quantum efficiencies of 62% were observed using a simple, untuned two-mirror cavity with an output coupling of 40%. More recent experiments have employed a Q-switched, frequency-doubled Nd:YAG laser as the pump source. The input-output energy curves obtained with the 532-nm pump appear in Fig. 8, for two different crystals, one doped with approximately 0.02% Ti, the other with 0.1% doping. The output from the $Ti:Al_2O_3$ laser was in the form of a single, gain-switched pulse, with a duration of 5 ns at high pumping levels.

In other experiments a $Ti:Al_2O_3$ laser cavity with two intracavity, Brewster-angle prisms was constructed to determine the $Ti:Al_2O_3$ laser tuning range. The tuning range observed was from 660-986 nm, limited to some extent by the finite bandwidth of the mirrors. Further experiments are in progress to extend the operating range further with alternate cavity optics.

Some preliminary work was done on cw pumping of a $Ti:Al_2O_3$ laser; the pump in this case was an argon-ion laser, operating on all lines in the blue-green wavelength region. True cw operation was observed with liquid-nitrogen cooling, used primarily to enhance the thermal conductivity of the Al_2O_3 host crystal; 70 mW of power at 750 nm was generated for 10 W of incident power. Quasi-cw operation with a threshold peak power of 9.8 W was observed with the crystal at room temperature by chopping the pump beam at a 16% duty cycle; a peak output power of 44 mW was obtained at 12.5 W of pump power. Calculations show that cw operation from the $Ti:Al_2O_3$ laser should be possible at room temperature with less than 1 W of pump power from an argon-ion laser, by improving the crystal heatsinking and use of an optical cavity with a smaller mode size.

The insertion loss in some of the $Ti:Al_2O_3$ laser crystals, particularly those with higher doping levels (0.1%), was found to be relatively high, on the order of 5%/cm. Absorption measurements by conventional techniques showed the presence of an additional absorption band in some of the

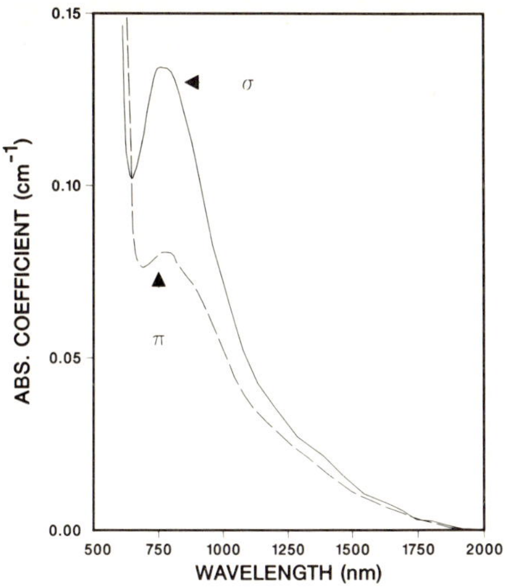

Fig. 9. Polarized long-wavelength absorption band in $Ti{:}Al_2O_3$

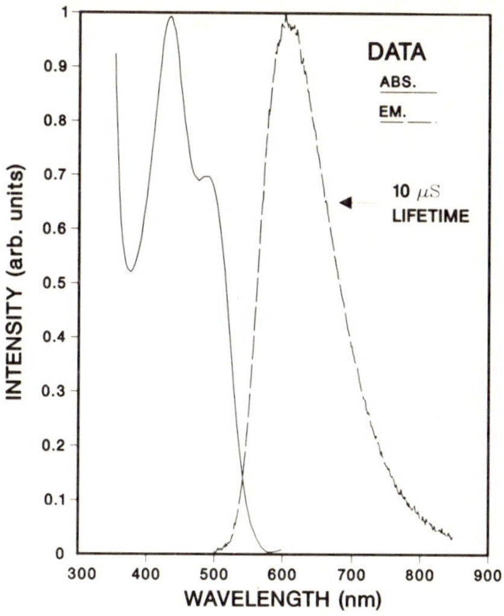

Fig. 10. Absorbance and emission spectra from $Ti{:}YAlO_3$ at room temperature

crystals tested; polarized absorption data are shown in Fig. 9, for a crystal with approximately 0.08% doping. The absorption is not associated with Ti^{3+} ions in the normal site positions, nor does it appear to be connected with pairs of Ti^{4+} and Fe^{2+} ions, since the spectra for such pairs [3] differs from the data of Fig. 9. The cause and elimination of the unwanted absorption are currently under investigation.

It is expected that with higher-quality crystals improvements in all aspects of the $Ti{:}Al_2O_3$ laser will be observed. Given the superior thermal and mechanical properties of the host crystal it is likely that the laser will be capable of high-average-power operation. The broad absorption band centered at 500 nm overlaps the output lines of the argon-ion, Cu-vapor and doubled Nd-laser systems to allow pumping by any of those sources. Flashlamp pumping is also possible, but for full energy storage in Q-switched operation lamp pulsewidths will have to be 3 μs or less. The gain cross section of the $Ti{:}Al_2O_3$ laser transition is sufficiently high to allow short-pulse Q-switched operation without excessive cavity fluences, unlike other tunable transition-metal lasers. Current work involves operation of the $Ti{:}Al_2O_3$ at high energies and a search for new host crystals for Ti^{3+}.

The absorption and emission spectra of $Ti{:}YAlO_3$, shown in Fig. 10, are an example of data on new hosts, and illustrate the sensitivity of the Ti^{3+} levels to the crystalline environment; in the case of the $YAlO_3$ host the $^2T_2$-^2E level separation is greater than in Al_2O_3 and both absorption and emission bands are blue-shifted. The emission lifetime in the samples examined to date is 10 μs, but the quantum efficiency is not yet known. There are undoubtedly a number of other crystals that can be grown doped with Ti^{3+}, and it is likely that hosts other than Al_2O_3 can be used for laser operation with the Ti^{3+} ion.

References:

1. L. F. Johnson, R. E. Dietz and H. J. Guggenheim, Appl. Phys. Lett. 5, 21 (1964).
2. L. F. Johnson, H. J. Guggenheim and R. A. Thomas, Phys. Rev. 149, 179 (1966).
3. J. Ferguson and P. E. Fielding, Chem. Phys. Lett. 10, 262 (1971).

Laser Action in Cr-Doped Garnets and Tungstates

Günter Huber and Klaus Petermann

Institut für Angewandte Physik, Universität Hamburg, Jungiusstraße 11
D-2000 Hamburg 36, Fed. Rep. of Germany

1. Introduction

Within the class of vibronic transition metal solid state lasers Cr^{3+}-doped crystals represent favourable candidates for room-temperature applications. Due to the high stability of the valency of Cr^{3+}, pure compounds with specific trivalent types of luminescent centers can be grown. Mostly high quantum efficiencies at room-temperature are observed.

Pulsed room-temperature lasing has been demonstrated in alexandrite [1] and Cr-doped gallium garnets [2,3]. The first tunable cw laser operation of Cr^{3+} was obtained in crystals belonging to the class of Ga-garnets, e. g. GSGG or GGG [4], and also in alexandrite [5].

In contrast to ruby, Cr-doped materials can also exhibit a broad band 4-level system for the $^4T_2$-$^4A_2$ transition, if a low crystal field at the Cr site and/or a large relaxation of the $^4T_2$ level yield a relatively small $^4T_2$-$^4A_2$ splitting (see Figs. 1,2). Thus, due to the thermal population of the $^4T_2$ level and the high $^4T_2 \rightarrow {}^4A_2$ transition probability, most of the fluorescence can be channelled into the broad band $^4T_2$-$^4A_2$ transition.

Recently further room-temperature lasers have been operated such as emerald [6], the Al-garnet Cr:GSAG [7], and Cr:$KZnF_3$ [8].

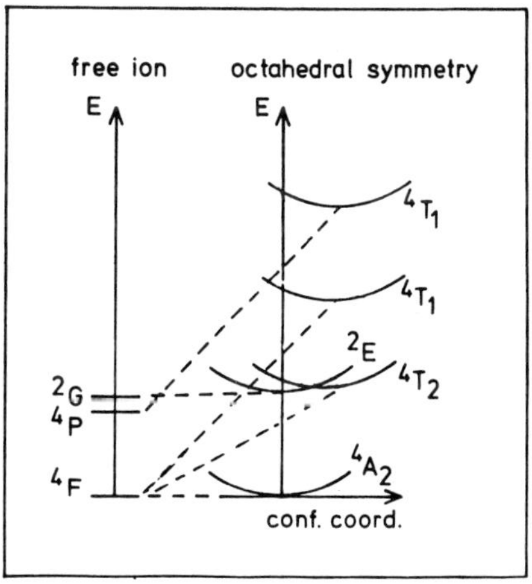

Fig. 1: Configurational diagram. Relevant free ion levels and resulting level scheme for octahedrally coordinated Cr^{3+}

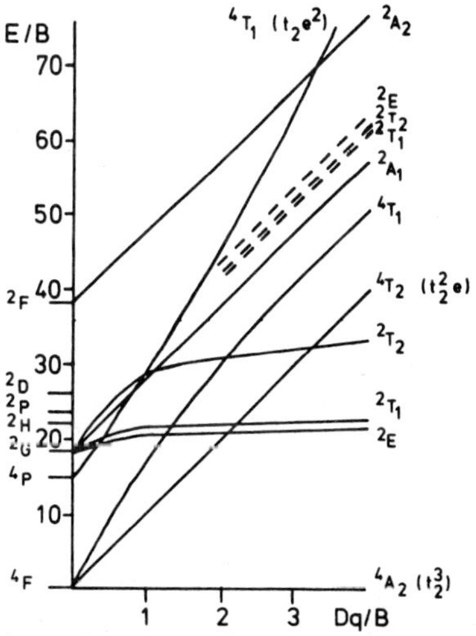

Fig. 2: Energy levels vs. crystal field strength in units of the Racah-parameter B for octahedrally coordinated Cr^{3+} [9]

This paper reviews the results obtained for Cr-doped Ga-and Al-garnets as well as Cr-doped tungstates.

2. Basic Spectroscopic Properties

The relative intensity of the R-lines and broad band emission depends on the energy gap

$$\Delta E = E(^4T_2) - E(^2E) \qquad (1)$$

in Figs. 1 and 2. For instance, ΔE is positive for alexandrite ($\sim$800 cm^{-1}) [1] and negative for Cr:$(La,Lu)_3(Lu,Ga)_2Ga_3O_{12}$ (-1000 cm^{-1}). Figure 3 shows the emission spectra of different Ga-garnets. With increasing lattice constant, the relative intensity of the R-lines decreases and the broad band fluorescence is shifted to longer wavelengths.

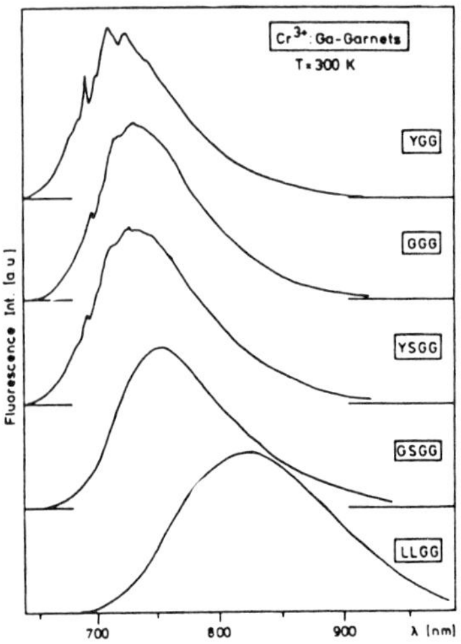

Fig. 3: Fluorescence spectra of Cr-doped Ga-garnets $Y_3Ga_5O_{12}$ (YGG), $Y_3(Sc,Ga)_2Ga_3O_{12}$(YSGG), $Gd_3Ga_5O_{12}$(GGG), $Gd_3(Sc,Ga)_2Ga_3O_{12}$(GSGG) and $(La,Lu)_3(Lu,Ga)_2Ga_3O_{12}$(LLGG) [4]

The weak crystal field can also be realized in Al-garnets. As an example, Figs. 4 and 5 show the absorption and fluorescence spectra of Cr:$Gd_3Sc_2Al_3O_{12}$(GSAG). Even at 4K the broad band emission dominates. It consists of the R_1-line (691.0nm) and the superposition of the broad band $^4T_2$-$^4A_2$ transition and the R_1-line phonon sidebands.

In Cr:$ZnWO_4$ (Fig. 6) the redshift of the fluorescence is even larger. This is due to the large Zn^{2+}-site and due to the low monoclinic site symmetry 2 which splits the $^4T_2$ level by 1640 cm^{-1}. Therefore the energy of the lower component of the $^4T_2$ level is decreased by both effects, site size and site symmetry.

The lifetime of the ^2E level is of the order of milliseconds, because the $^2E \rightarrow {}^4A_2$ transition is parity and spin forbidden. The radiative $^4T_2$ lifetime is one to two orders of magnitude shorter depending on the actual site symmetry of the Cr ion. In Cr:$ZnWO_4$ the acentric site forces electrical dipole transitions (radiative $^4T_2$-lifetime 8.7 μs), whereas in Cr:GSGG the dynamical distortion of the centric site $\bar{3}$ by acentric lattice vibrations yields a radiative $^4T_2$-lifetime of about 70 μs (extrapolation from LLGG).

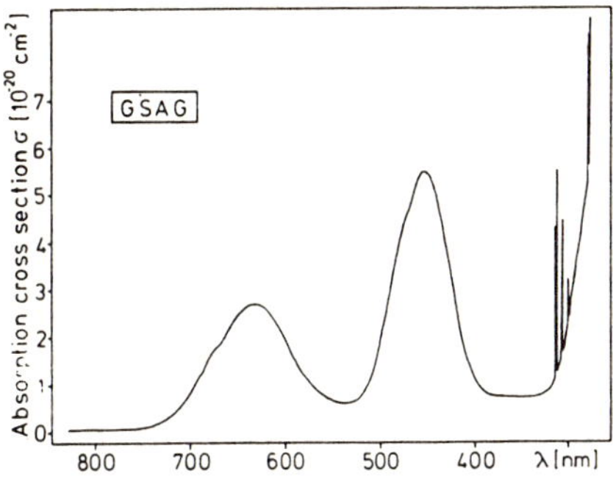

Fig. 4: Absorption spectrum of $Cr:Gd_3Sc_2Al_3O_{12}$(GSAG) [7]

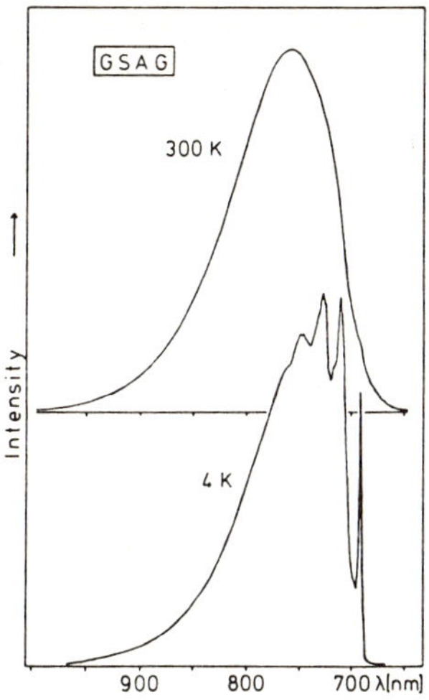

Fig. 5: Fluorescence spectra of $Cr:Gd_3Sc_2Al_3O_{12}$(GSAG) at 4K and 300K [7]

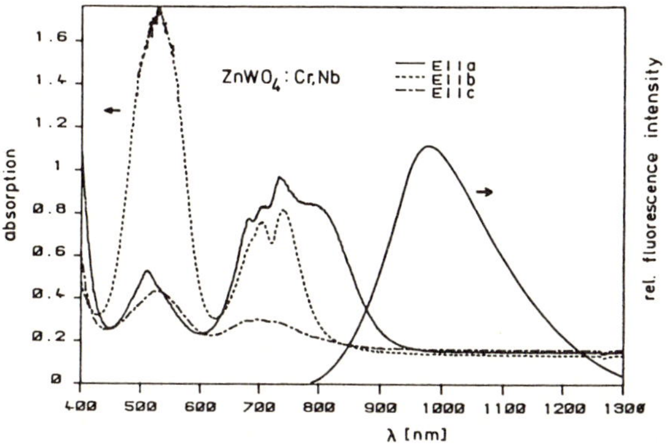

Fig. 6: Absorption and fluorescence spectra of $Cr:ZnWO_4$ [10]

The measured decay curves in $Cr:ZnWO_4$ strongly depend on the actual charge compensation in the crystals. Typical decay curves are shown in Fig. 7 for the case of pentavalent Nb compensation [10]. The curve at 77K represents the radiative decay of $Cr:ZnWO_4$ crystals with intrinsic charge compensation (lifetime τ = 8.7 µs) as well as doubly doped Cr, $Nb:ZnWO_4$ crystals with extrinsic charge compensation. In general, codoping of monovalent and pentavalent ions decreases the quantum efficiency. The highest value of 30 % was found for crystals without any codoping, i. e., in this case charge compensation is realized intrinsically by Zn-vacancies [13]. Due to the twofold valency of the Zn-vacancy two types of Cr^{3+}-centers are created with a 1:1 ratio:A high quantum efficiency center (τ(77K) = 8.7 µs, τ(300K) = 5.4 µs) with an adjacent Zn-vacancy, and a strongly quenched center (τ(77K) = 8.7 µs, τ(300K) = 0.5 µs) with a non-local compensation [13]. The Cr doped garnets show totally exponential decays (see Fig. 8).

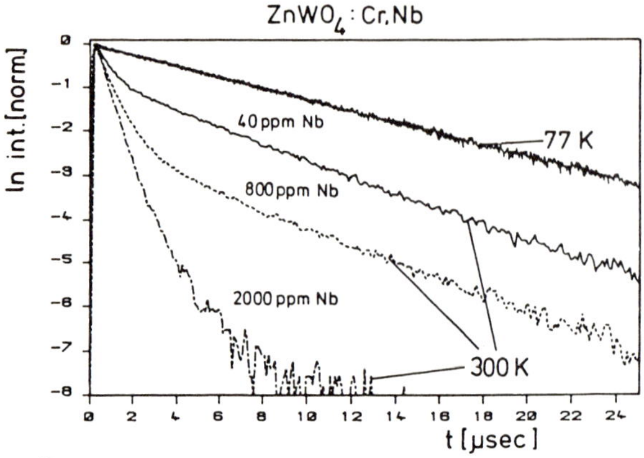

Fig. 7: Decay curves of Cr^{3+}:$ZnWO_4$ showing the dependence on Nb charge compensation and the radiative lifetime of 8.7 μs at 77K

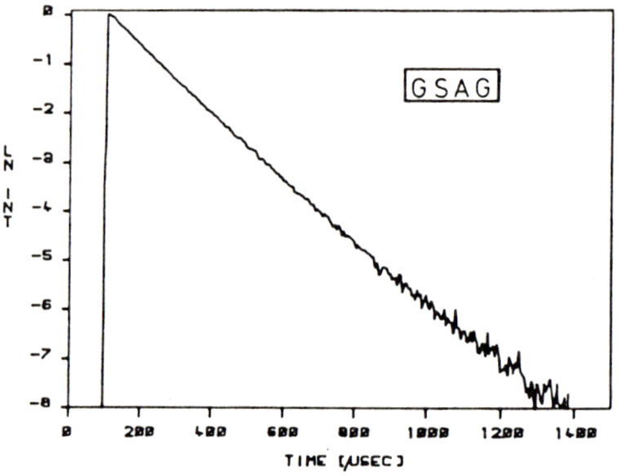

Fig. 8: Log-plot of the exponential decay of Cr:GSAG [7]. Lifetime is $\tau = 150$ μs

The spectroscopic parameters of both, Cr-doped garnets and tungstates are summarized in Table 1. The lifetime τ decreases with increasing site size. This is mainly due to the decreasing energy gap ΔE which is negative for LLGG and $ZnWO_4$. As expected the Racah parameter B is nearly constant. The relative crystal field strength Dq/B decreases from YGG to LLGG by about 10 %. The effective emission cross sections σ_e are calculated from the spectroscopic data using the theory of Mc Cumber [11]. The variation within the garnets arises from the ΔE-dependence of the Boltzmann occupation of the $^4T_2$ level. The high cross section and the short lifetime in Cr:$ZnWO_4$ is caused by the acentric Cr site. The two values for τ and ΔE correspond to the low and high quantum efficiency center, respectively.

Table 1: Basic spectroscopic parameters of Cr-doped garnets and $ZnWO_4$ at 300 K

Host	YGG	GGG	GSAG	YSGG	GSGG	LLGG	$ZnWO_4$
τ [μs]	241	159	150	139	115	68	0.5/5.4
ΔE [cm^{-1}]	650	380		350	50	-1000	-2700/-2100
B [cm^{-1}]	656	645		650	658	651	600
Dq/B	2.30	2.28		2.27	2.20	2.07	2.05
σ_e[$10^{-20}cm^2$]	0.36	0.6		0.6	0.8	1.6	40...50
λ_o [nm]	730	745	770	750	770	850	1030

3. Laser Experiments

The cw laser performance of the Cr-crystals has been measured by laser exitation with an argon or krypton ion laser. The typical experimental set-up is shown in Fig. 9. The tuning element is a Brewster-angle prism or a two-plate-birefringent filter. The laser wavelength can simultaneously be monitored with a spectrometer. The laser crystal is placed directly in front of the nearly plane (r=4m) input mirror. The pump beam is focused by a lens (f=5cm to 10cm) onto the sample.

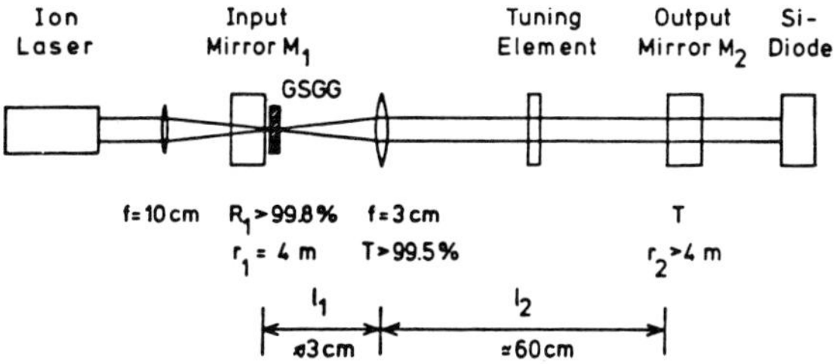

Fig. 9: Typical resonator for testing laser action in Cr-doped crystals

3.1 Cr-Doped Garnets

From the garnets YGG, GGG, YSGG, GSAG, GSGG and LLGG laser action at room temperature could be obtained. With respect to the gain GSAG, GSGG and LLGG are favourable compounds. Figure 10 compares the cross sections of several garnets. Principally, in LLGG the highest gain can be achieved.

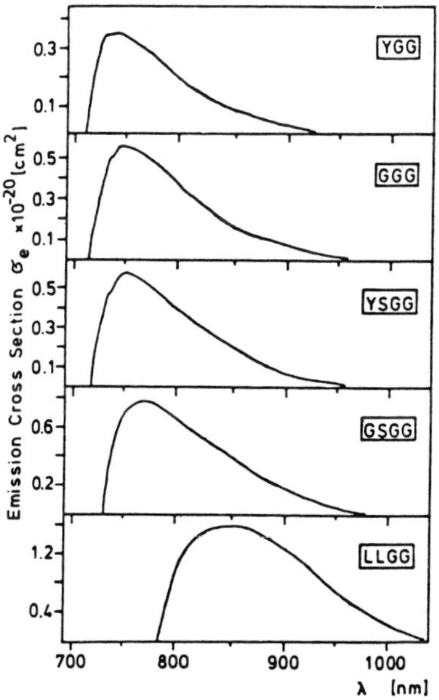

Fig. 10: Effective cross sections σ_e of the laser transition $^4T_2 \rightarrow {}^4A_2$ for several garnets [15]

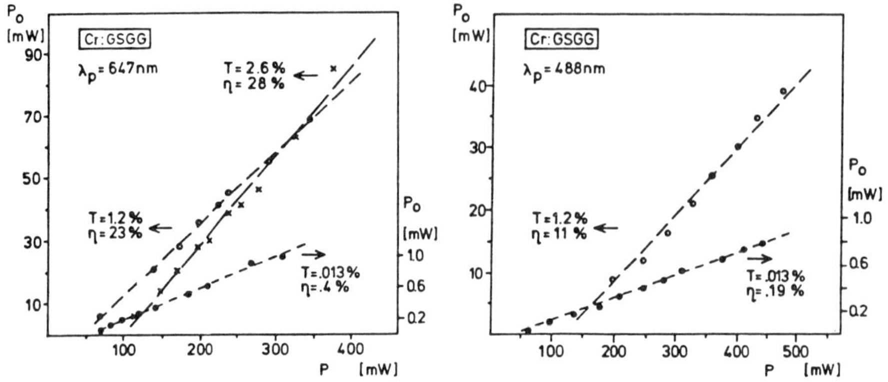

Fig. 11: Laser output P_o of Cr:GSGG for different output couplings and pump wavelengths λ_p = 647 nm (a) and λ_p = 488 nm (b). Cr-concentration $n_{cr} = 1\cdot 10^{20}\ cm^{-3}$ [12]

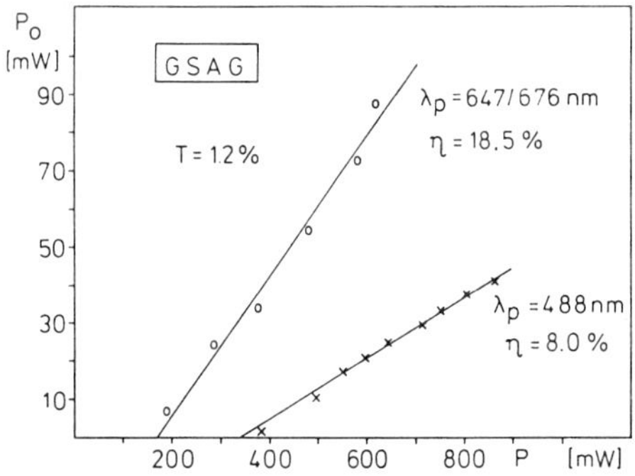

Fig. 12: Performance of the Cr:GSAG laser at different pump wavelengths λ_p. Output coupling T = 1.2 %, Cr-concentration $n_{cr} = 5.7\cdot 10^{19}\ cm^{-3}$ [7]

Detailed laser experiments were made with a krypton pump laser at λ_p = 647/676 nm and an argon laser at λ_p = 488 nm. The input/output performance of Cr:GSGG and Cr:GSAG is shown in Figs. 11 and 12, respectively.

In Cr:GGG (λ=769 nm, $n_{cr}=4\cdot 10^{19} cm^{-3}$) the slope efficiency η is 10 % and the pump threshold is 240 mW for T = 1.2 %. In Cr:LLGG (λ = 830 nm, $n_{cr}=6\cdot 10^{19}\ cm^{-3}$) a fading process is observed [12], which limits the slope efficiency to 3 %. However, this fading behaviour, which appears within the first ms of pumping, is probably caused by a colour center formation due to impurities in the starting materials for the crystal growth. In YGG and YSGG, the small effective cross sections yield high inversion densities at the pump threshold. The observed slope efficiencies are about 5 %.

A detailed analysis of Fig. 11 is given in reference [12]. The Cr-GSGG laser is less efficient at λ_p = 488 nm. This effect can be explained by excited state absorption of the pump light. Introducing the pump efficiency ε the laser losses L can be determined from

$$\eta = \frac{\lambda_p}{\lambda} \cdot \varepsilon \cdot \frac{T}{T+L} . \qquad (2)$$

The resulting value is L = 2.7 % $\pm$ 0.3 % for ε(647 nm) = 0.83 and ε(488 nm) = 0.65. These data are confirmed by the relation of the threshold condition and the inversion density N_{th}

$$2N_{th}\,\sigma_e\, l = T+L, \quad (3)$$

where l is the crystal length and σ_e the effective cross section of Table 1.

3.2 Cr:$ZnWO_4$

The doping level of Cr in $ZnWO_4$ is typically 1000 ppm ($n_{Cr}=1.5\cdot 10^{19}$ cm^{-3}). The maximum absorption cross section into the $^4T_2$ pump level is $6\cdot 10^{-19} cm^2$. Therefore typical crystal lengths are of the order of mm as for the garnets. The calculated 77K-gain of a 800 ppm sample at 5 % inversion is shown in Fig. 13. At the short wavelength side the influence of the ground state absorption limits the minimum wavelength to about 890 nm.

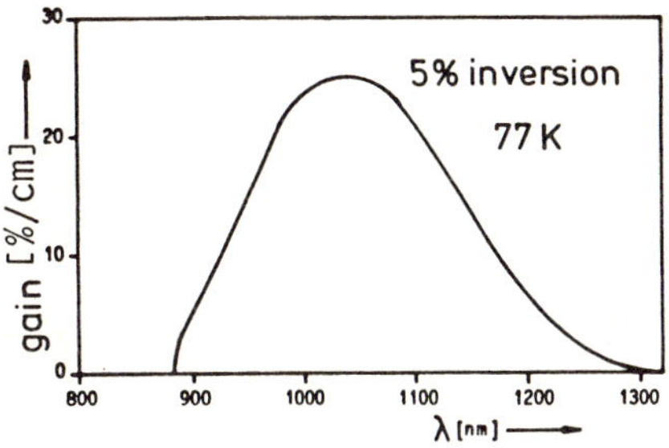

Fig. 13: Cr:$ZnWO_4$. Gain curve of a 800 ppm sample at 5 % inversion. Maximum emission cross section is $4.3\cdot 10^{-19}$ cm^2 [13]

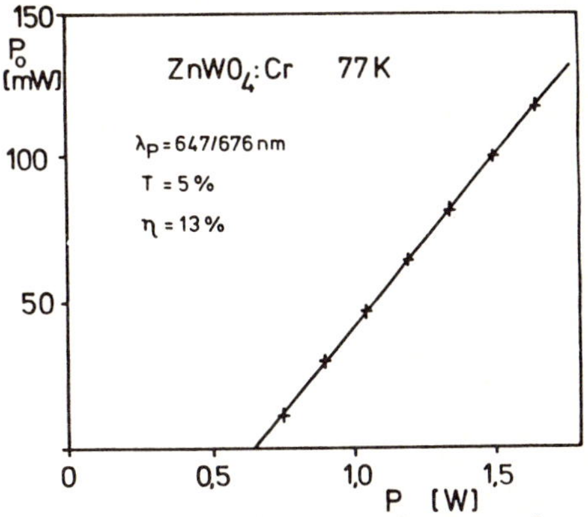

Fig. 14: Cw laser perfomance of Cr:$ZnWO_4$ at 77K. Longitudinal excitation with a krypton laser beam at 647/676 nm [13]

As expected from the fluorescence and the gain the free running cw laser operates near 1 μm wavelength with E||a. The input/output data are shown in Fig. 14. Cw operation was not possible at 300 K. This is due to the low quantum efficiency (~ 30 %) and the poor thermal conductivity. The improvement of these factors is under investigation. The lossy cavity and the high gain of Cr:$ZnWO_4$ have prevented smooth and stable tuning of the wavelength by a prism or birefringent filter. With different mirror coatings it was possible to operate the laser at various wavelengths between 980 nm and 1090 nm.

3.3 Tunability of the Cr:GSGG Laser

The tuning range in Fig. 15 with maximum output of 200 mW was obtained with one output mirror (T = 2.5 %...1.4 % in the range 740 nm...850 nm).

In order to get the whole tuning range a high reflectivity cavity ($R_1=R_2>99.8\%$) was used for further experiments. Figure 16 shows the wavelength dependence of the pump power at threshold P_{th} from 742 nm to 842 nm for the high reflectivity set-up. The threshold power P_{th} (λ) can be used to calculate $\sigma_e(\lambda)$ by (3). Figure 17 compares the resulting cross sections with those obtained from the fluorescence and absorption spectra. Note, that the ground state absorption is already involved in the curve, which

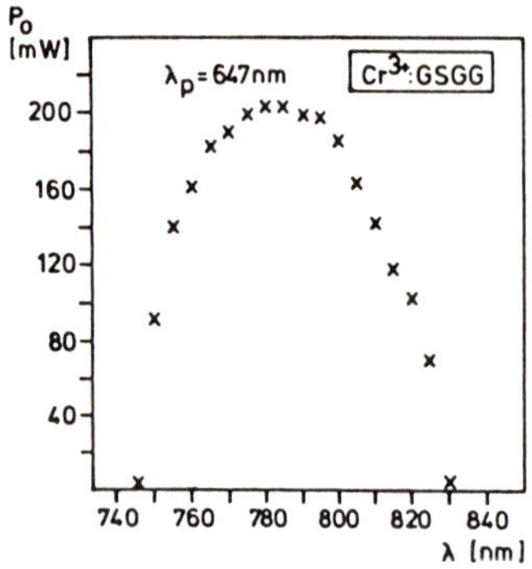

Fig. 15: Tunable output P_o of the Cr:GSGG laser. The absorbed pump power is 1.1 W [12]

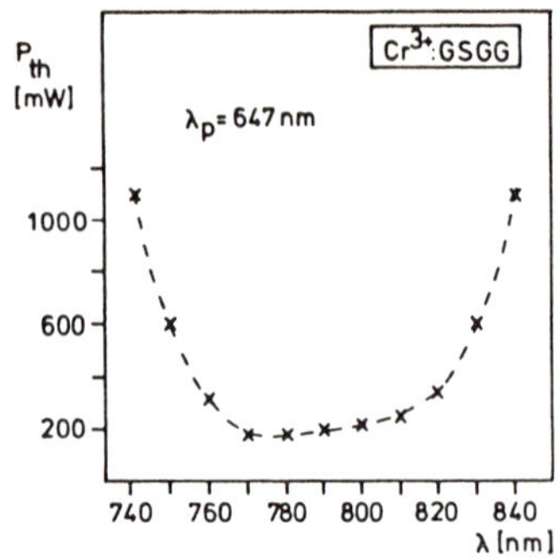

Fig. 16: Wavelength dependence of the threshold pump power P_{th} (λ) in a high reflectivity cavity [12]

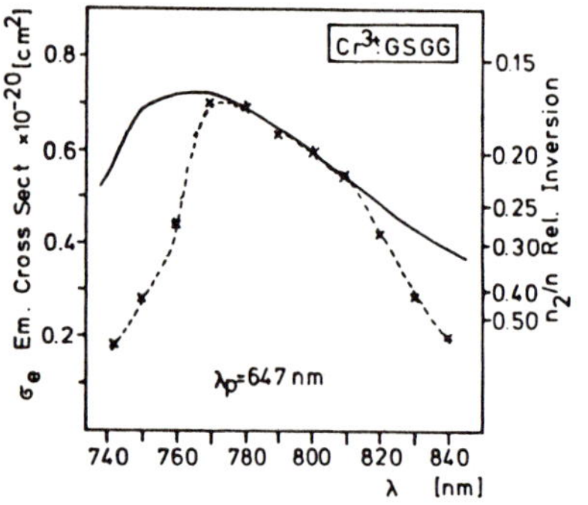

Fig. 17: Comparison of the effective cross sections σ_e (λ) obtained from the fluorescence and absorption spectra (solid line) and the threshold pump power (dashed line) [12]

is related to the fluorescence spectrum. The steep decrease of $\sigma_e(\lambda)$ at short wavelengths is ascribed to the onset of excited state absorption of laser photons.

This fact was proved by measuring the excited state absorption directly [14]. The result is shown in Fig. 18, where I_u denotes the intensity of the test beam without pump beam and I_p denotes the intensity of the test beam with simultaneous krypton laser pumping at λ_p = 647 nm. The onset of excited state absorption σ_2 is at longer wavelength than the ground state absorption σ_1 and cancels the emission cross section near 750 nm.

The deviation in Fig. 17 at longer wavelength may be caused by the high average inversion density which limits the gain by deforming the gain profile by increased excited state absorption of pump photons.

Conclusion

Cr^{3+}-doped garnets and tungstates represent broad band 4-level systems which can be applied to generate tunable coherent radiation in the red and near infrared spectral region. $Cr:ZnWO_4$ is the first example of a tunable Cr^{3+}-laser beyond 1 µm. The "expanded" Cr-doped garnets can be applied as tunable room temperature lasers in the spectral region between 740 nm and 840 nm (Cr:GSGG) and even longer wavelengths (Cr:LLGG). Tunable

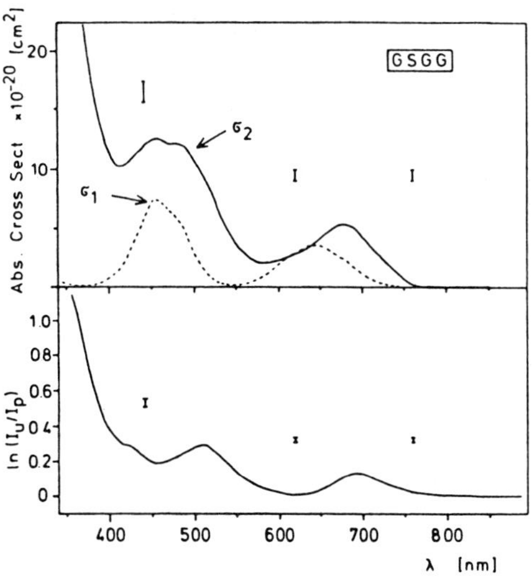

Fig. 18: Measurement of excited state absorption. I_u and I_p denote the intensity of the test beam without and with krypton laser pumping at λ_p = 647 nm, respectively [15]. σ_1 = ground state absorption, σ_2 = excited state absorption

cw output up to 200 mW has been achieved in Cr:GSGG. In the Al-garnet hosts like GSAG similar laser properties were found.

References

1. J. C. Walling, O. G. Peterson, H. P. Jenssen, F. C. Morris, E. W. O'Dell: IEEE J. QE-16, 1302 (1980)
2. B. Struve, G. Huber, V. V. Laptev, I. A. Shcherbakov, E. V. Zharikov: Appl. Phys. B 28, 235 (1982)
3. E. V. Zharikov, N. N. Il'ichev, S. P. Kalitin, V. V. Laptev, A. A. Malyutin, V. V. Osiko, V. G. Ostroumov, P. P. Pashinin, A. M. Prokhorov, V. A. Smirnov, A. F. Umyskov, I. A. Shcherbakov: Sov. J. Quantum Electron. 13 (9), 1274 (1983)
4. B. Struve, G. Huber, V. V. Laptev, I. A. Shcherbakov, E. V. Zharikov: Appl. Phys. B 30, 117 (1983)
5. S. T. Lai, M. L. Shand: J. Appl. Phys. 54, 5642 (1983)
6. M. L. Shand, S. T. Lai: IEEE J. QE-20, 105 (1984)
7. J. Drube, B. Struve, G. Huber: Optics Commun. 50, 45 (1984)
8. U. Brauch, U. Dürr: Optics Commun. 49, 61 (1984)
9. Y. Tanabe, S. Sugano: J. Phys. Soc. (Japan) 9, 766 (1954)
10. G. Huber, B. Struve, K. Petermann, W. Kolbe: Proc. Internat. Conf. Lasers '82, New Orleans, Louisiana (USA), STS Press, Mc Lean, 1983
11. D. E. Mc Cumber: Phys. Rev. 134, 299 (1964)
12. B. Struve, G. Huber: J. Appl. Physics, to be published
13. W. Kolbe, K. Petermann, G. Huber: CLEO '84, paper WI1, Anaheim (USA), 1984
14. B. Struve, G. Huber: CLEO '84, paper WI2, Anaheim (USA), 1984
15. B. Struve, G. Huber: to be published

Optical and Lasing Properties of Cr^{3+}, Co^{2+}, Ni^{2+} and V^{2+} Doped Perovskites

U. Dürr, U. Brauch, W. Knierim, and C. Schiller

Universität Stuttgart, Phys. Institut Teil 2, Pfaffenwaldring 57
D-7000 Stuttgart 80, Fed. Rep. of Germany

1 Introduction

In the last years insulating single crystals doped with transition metal ions (TM) have proved their potential as laser material for laser systems tunable in the infrared spectral region. Since the first investigations of the laser properties of such so called vibronic lasers [1] several systems have successfully operated cw even at room temperature [2-6]. In this paper we survey the electronic and lasing properties of TM in alkali-metal-halides. The goal of these investigations is the development of vibronic lasers on the basis of these materials which should have properties as close as possible to the following requirements:

a) these lasers should be broadly tunable so that few systems are necessary to cover the infrared spectral region. Therefore, reabsorption and excited state absorption should be negligible.

b) the lasers should operate cw at room temperature at a high output level.

c) the lasers should have a high total efficiency which means that a high quantum efficiency and broad pump levels are desirable.

d) last but not least the laser material should be optically and mechanically stable and isotropic. The growth of large single crystals should be possible. The crystals also must have a reasonable room temperature heat conductivity and little nonlinear behaviour.

In the first part of this paper some of the requirements mentioned above will be discussed in more detail to give the reader a feeling how to select a reasonable laser material for a desired spectral region and also to point out the spectral limits for laser operation with this material. In the second part, a short review is given about the already lasing TM doped fluorine perovskites with emphasis on the $KZnF_3:Cr^{3+}$ vibronic laser, which may come close to the above mentioned requirements.

2 The Host Material

Some short remarks should be given about the host material. In chosing alkali(A)-metal(Me)-halides(X) doped with 3d, 4d or 5d transition metals one is confronted with a vast number of possible chemical compositions and structures [7] like the simple cubic $AMeX_3$ with perovskite structure where the Me-ion is surrounded by an octahedron of halogen ions (see insert in fig. 1) More complicated systems are the elpasolites (A_2AMeX_6) or the garnets

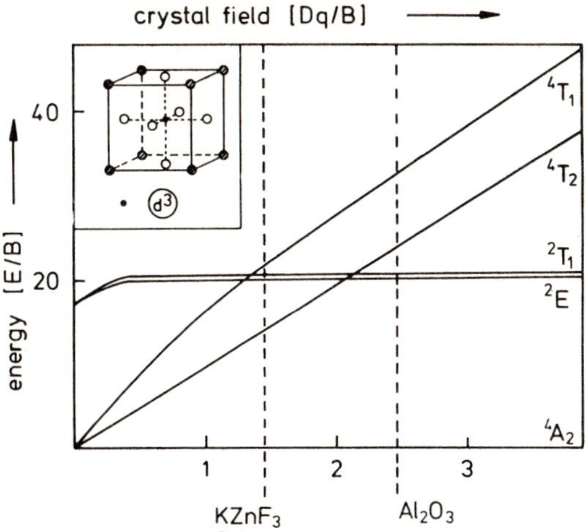

Fig.1 Reduced Tanabe-Sugano diagram for d^3 transition metal ions in octahedral site symmetry

($A_3A_3'Me_2X_{12}$).Only fluorine single crystals and few chlorine compounds offer reasonable material properties as requested in the introduction.In this paper we restrict the discussion mainly on the 3d TM in $AMeF_3$host crystals with cubic symmetry like $KMgF_3$ or $KZnF_3$

3 Spectral Range of the Laser Transitions

The first question discussed is: which spectral range may be covered by these systems?Simple crystal field theory which describes in good approximation the electronic levels of TM in ionic crystals [8] will give the answer.All information can be deduced from the so called Tanabe-Sugano (TS) diagram which gives the shift and splitting of the electronic levels in dependence of the strength of the cubic crystal field.Figure 1 shows the reduced TS diagram for d^3 ions like V^{2+},Cr^{3+} or Mn^{4+}.The only level of interest in the following is the ^4F groundstate of the free d^3 ion whose orbital degeneracy is partly lifted in the octahedral electric crystal field from the ligands.The two lowest levels are the $^4A_2$ groundstate and the $^4T_2$ excited state for crystal fields smaller than 2.These electronic states are differently coupled to the host lattice.Under this condition phonon-assisted optical transitions are expected which lead to broad vibronic sidebands in absorption and emission.In simple crystal field theory the splitting of these levels depend on the crystal field parameters D,q and the Racah parameter B.The proper selection of these parameters should enable one to select the wavelength range of transition between the two lowest levels which are the laser levels:

a) $q \sim \int |R(3d)|^2 r^6 dr$, (R(3d) is the radial part of the wavefunction of the 3d TM)
Different $3d^n (n \neq 5)$ TM should show different crystal field splittings of the groundstate even in the same host lattice. This can be clearly seen in Fig.2

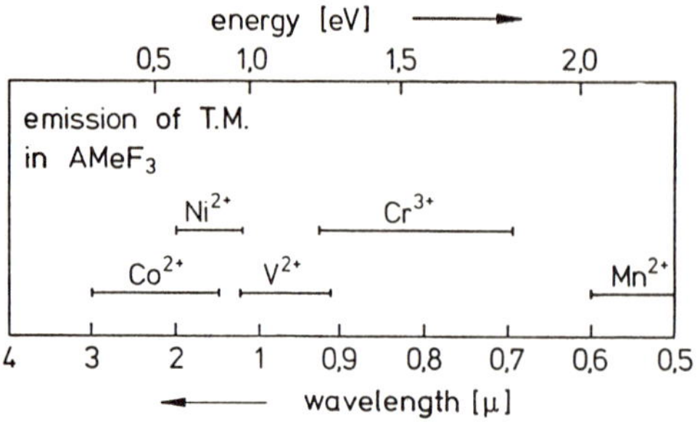

Fig.2 Spectral range of emission for different transition metal ions in cubic $AMeF_3$. Emission occurs between the two lowest levels

b) $D \sim a^{-5}$, (a is the lattice constant of the cubic crystals) Detailed investigations on Co^{2+} doped $AMeF_3$ have shown that the dependence expected from theory is only observed when host crystals with different alkaline ions are compared. The influence of the change in lattice constant is less than expected when crystals with different Me-ions are compared [9]. This is probably due to local distortions at the impurity site and to the influence of covalence. Similar observations have been made with d^3 ions shown in Fig.3

c) The Racah parameter B represents the intraion electron-electron interaction. B decreases with increasing ionic radius as can be seen from the B values for the isoelectronic free ions V^{2+} and Cr^{3+} : Cr^{3+} (r=0.76Å, B=918 cm^{-1}) V^{2+} (r=0.98Å, B=755 cm^{-1}) [8]. The influence on B also is manifest in Fig.3. The splitting of the laser levels in V^{2+} doped perovskites is typically 0.2μ shifted to longer wavelength compared to Cr^{3+} in similar hosts.

d) Not included in the simple crystal field theory is the influence of the wavefunction of the ligands. This influence is reflected in the well known spectrochemical series which predicts, on the basis of numerous spectroscopical data, that the splitting of the TM levels increases going from Cl^- to F^- and O^- ligands. The validity of this empirical rule is also evident in Fig.3 and Fig.1.

On the basis of these considerations the splitting of the levels for TM in suitable host materials may be estimated. The conclusion is that the spectral range between 5μ -0.5μ may be covered by these systems. Up to now the emission experiments show a possible range between 3μ -0.7μ for TM in $AMeX_3$ (Fig.4) when only fluorides and chlorides are considered [10] .

4 Nonradiative Transitions between Laser Levels

An important parameter when dealing with laser material is the quantum efficiency for emission and its temperature dependence. As an example, the room temperature values for the d^3 TM are inserted in Fig.3. These numbers make clear that not all systems are suitable as laser material for room temperature operation, some even not at low temperatures.. One possible reason is the site symmetry at the TM impurity which is octahedral for most of the $AMeF_3$ but trigonal for the $AMeCl_3$ shown in Fig.3. This argument will be discussed in more detail in what follows.

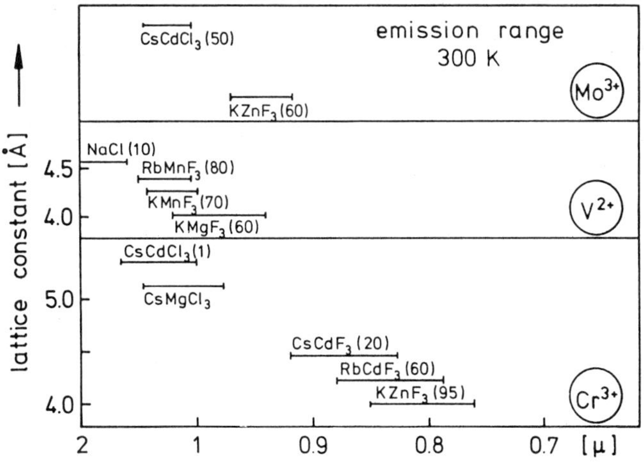

Fig.3 Spectral range of emission for d^3 ions in host materials with different lattice constants and ligands.The inserted numbers give the room temperature quantum efficiency for emission.

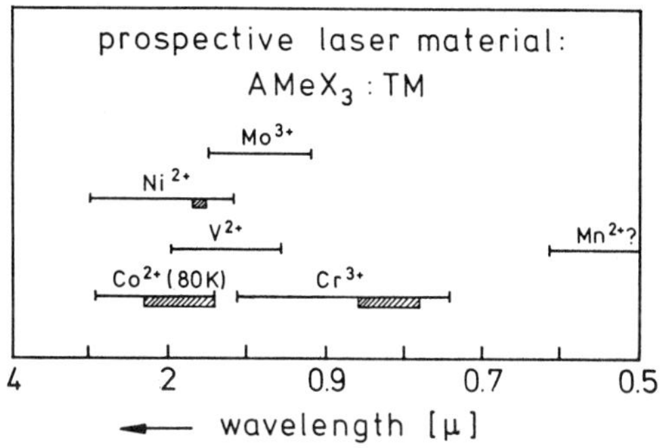

Fig.4 Spectral range of possible laser emission of TM in $AMeX_3$ (X=F,Cl). The hatched bars give the tuning range of actual lasers

There are several physical mechanisms which may be responsible for non-radiative transitions between electronic levels [11]. We assume that the electron-phonon coupling is mainly responsible for the relaxation processes in TM doped perovskites.The two different relaxation channels are indicated in Fig.5.Here the groundstate Ψ_i with symmetry Γ_i and the first excited state $\Psi_f(\Gamma_f)$ are drawn in a configuration coordinate diagram.After optical excitation of the defect a fast relaxation takes place to the lower vibronic levels in the excited state.In this process S phonons of the average energy $\hbar\Omega$ are emitted from the local electronic defect into the lattice.S is dependent on the strength of the electron-phonon coupling.For the systems discussed in this paper S is typically between 2 and 4 and indicates a weak coupling of the laser levels to the lattice.The consequence is that the relative shift of the two parabolas in the configuration diagram is small.Calculations show [12] that in our systems the energy of the crossing point is typically more than $E_A = 2\cdot 10^4$ cm^{-1} (see Fig.5)..The probability to reach this point with thermally excited

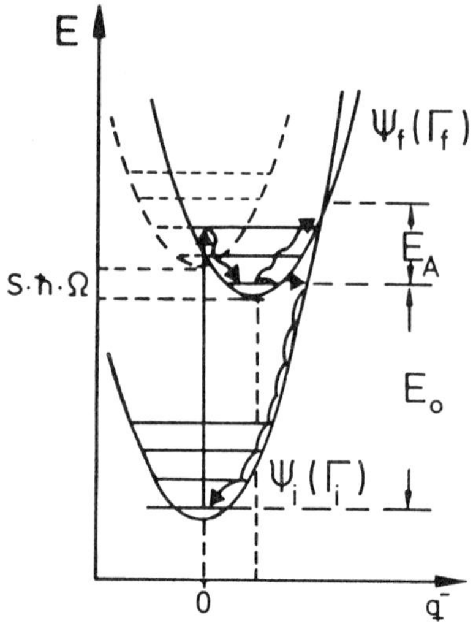

Fig.5 Relaxation channels in an electronic system coupled to phonons. E_O is the energy of the zero phonon transition. E_A is the energy of the crossing of the parabolas above the vibronic groundstate of the excited state (details see text).

phonons is small.Therefore we believe that the tunneling process between the excited state and the groundstate is the main relaxation channel for the TM laser centers .In the tunneling process the two electronic states are coupled by the "promoting" phonons. The excitation is transfered to the "accepting" optical phonons in the groundstate and carried away from the electronic defect by the acoustic phonons.The probability for this process is a complicated function of S,the Energy E_O of the zero phonon transition, the frequency of the promoting mode and the average number of accepting modes [10] .The probability also depends on the tunneling matrix element $<\Gamma_i|\Gamma$(promoting mode$|\Gamma_f>$ which will be discussed in more detail.Apparently the criterion for the tunneling transition to take place is the existence of the promoting phonon mode of the proper symmetry at the $(MeX_6)^{3-}$ cluster which couples the electronic states.Simple group theoretical analysis of the transition matrix element for different TM ions in octahedral site symmetry shows that in this model the laser levels of d^3 and d^8 ions are not coupled,while for d^7 ions strong radiationless transitions are expected.This is indeed qualitatively observed:$Cr^{3+}(d^3)$ or $Ni^{2+}(d^8)$ ions in cubic $KZnF_3$ have a low temperature quantum efficiency of almost 100%,only about 30% have been measured for $KZnF_3:Co^{2+}$. Similar symmetry considerations make understandable the small quantum efficiency of Cr^{3+} in crystals with lower symmetry which is the case for $CsCdCl_3$ in Fig.3.

The consequence of these qualitative arguments is that one should choose cubic host crystals with octahedral local symmetry as laser material.When these crystals are substitutionally doped with d^3 and d^8 ions reasonable room temperature threshold and moderate thermal problems are expected.

5 Lasing Properties

Figure 4 shows the emission range covered by TM doped fluorine and chlorine perovskites suitable as laser material.The hatched bars indicate the tuning range of systems which already have

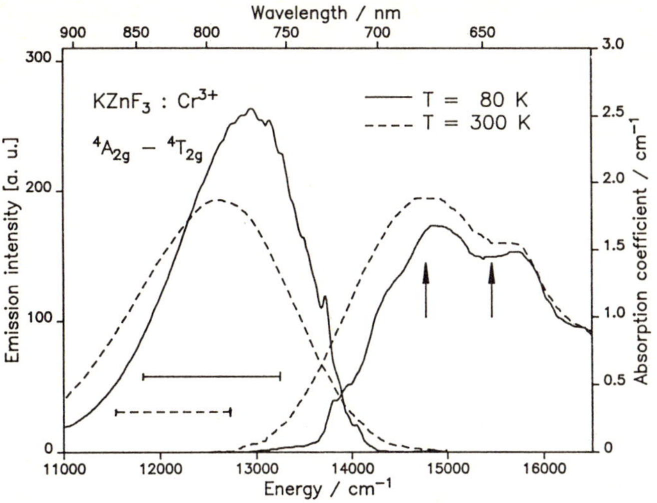

Fig.6 Absorption and emission spectra of the laser levels of $KZnF_3:Cr^{3+}$.The vertical arrows indicate the wavelength of the krypton laser pump.The horizontal bars give the tuning range at 80 K and RT [9,12]

shown cw laser output at low temperature or room temperature.These systems are $KMgF_3:Co^{2+}$,$KZnF_3:Co^{2+}$,$KMgF_3:Ni^{2+}$ and $KZnF_3:Cr^{3+}$.Some important properties of these lasers are collected in Table 1. In the following the laser properties of $KZnF_3:Cr^{3+}$ are shortly discussed and compared with other vibronic lasers which also operate at room temperature.

The absorption and emission of Cr^{3+} in $KZnF_3$ which takes place between the $^4A_2$ - $^4T_2$ levels is shown in Fig.6.The spectra are drawn for nitrogen temperature and room temperature.The crystals are optically pumped with the 647 nm or(and) the 676 nm lines of a krypton laser.At low temperatures (T<100K) the quantum efficiency for emission is almost 100% and decreases to room temperature to about (85±5)%.Typically the samples were doped with 0.5% Cr^{3+}. The samples which were placed in astigmatically-compensated three mirror resonators [12,13] were about 2.4 mm thick and absorbed 30% of the pump power.In the resonators,which differ in the curvature of the mirrors,different values for the threshold absorbed power and the slope efficiency have been observed.In Table 1 the values are given for the laser experiments,where the highest slope efficiency was observed..It is not quite clear why the slope efficiency for $KZnF_3:Cr^{3+}$ is less compared to the oxide systems presented in Table 2.This question is left to further investigations.Nevertheless the tuning range of this material is attractive.

First experiments with single mode operation of the $KZnF_3:Co^{2+}$ and $KZnF_3:Cr^{3+}$ lasers have demonstrated that this can be achieved relatively easily with frequency selective elements and etalons in the resonator.Up to now the linewidth given in Table 1 is instrument-limited.

Table 1 Laser Data for Transition Metal Ions Doped in Cubic Perovskites

	operat. temp. [K]	tuning range [μ]	slope eff. [%]	threshold [mW]	single mode operation
$KZnF_3:Co^{2+}$	80	2.18 - 1.65	31	<60	<80 MHz [18]
$KMgF_3:Co^{2+}$	80	1.9 - 1.62	20	<100	
$KMgF_3:Ni^{2+}$ [16]	77	1.591			
$KZnF_3:Cr^{3+}$	80 RT	0.845 - 0.758 0.865 - 0.785	14	<370	<1 GHz [17]

Table 2 Comparison of the RT properties of fluoride and oxide Cr^{3+} laser materials.All systems were pumped with a krypton laser

	tuning range nm	free laser nm	threshold mW	slope eff. %	heat conductivity mW/K·cm
alexandrite [14]	726-802	752	400	51	230
emerald [15]	729-809	765	500	34	56
GSGG [6]	742-842	777	230	28	90
$KZnF_3$	785-865	825	370	14	60

6 Summary

Theoretical considerations and the analysis of the experimental data of the optical properties of transition metal ions doped in alkali-metal-halides lead to the assumption that the spectral region between 0.7-2μ may be covered by vibronic lasers operating at room temperature.Above 2μ probably laser crystals on the basis of this material have to be cooled.The spectral limits are expected around 4μ due to the properties of the host crystals and the crystal field properties of the transition metal impurities in ionic crystals.

Acknowledgement
The authors gratefully acknowledge financial support by the Deutsche Forschungsgemeinschaft.

1 L.F.Johnson,H.J.Guggenheim and R.A.Thomas:Phys.Rev.149,179 (1966)
2 P.F.Moulton:Laser Focus May 1983,pp83
P.F.Moulton:IEEE Journal of Quantum Electr.,QE 18(8) ,1185 (1982)
3 U.Dürr:Laser und Optoelektronik 15,31 (1983)

4 J.Walling,O.G.Peterson,H.P.Jenssen,R.C.Morris and E.W.O'Dell: IEEE Journal of Quantum Electr.,QE 16(12),1302 (1980)
5 J.Buchert and R.R.Alfano:Laser Focus Sept.1983,pp117
7 Landoldt-Börnstein,New Series Vol.7a (Springer 1973)
8 S.Sugano,Y.Tanabe,H.Kamimura:"Multiplets of Transition Metal Ions in Crystals",Pure and Applied Physics Vol.33 (1970). (Wiley Interscience)
9 U.Dürr,U.Brauch,W.Knierim and B.Weigand:"Vibronic solid state lasers",Proc.Int.Conf.Laser 83,to be published
10 R.Englman:"Nonradiative decay of ions and molecules in solids" North Holland (1979)
11 C.Schiller:Diploma work (1984),unpublished
12 U.Brauch and U.Dürr:Opt.Lett.,submitted for publ.
13 W.Künzel,W.Knierim and U.Dürr:Opt.Comm.36(5),383 (1981)
14 M.L.Shand and S.T.Lai:IEEE Journal of Quantum Electr.QE 20(2) 105 (1984)
15 S.T.Lai and M.L.Shand:J.Appl.Phys.54(10),5642 (1983)
16 L.F.Johnson,H.J.Guggenheim,D.Bahnck and A.M.Johnson:Opt.Lett. 8(7),371 (1983)
17 cooperation with Dr.Litfin and coworkers (Spindler und Hoyer)
18 cooperation with Dr.German and coworkers (Burleigh Inst.)

Recent Progress in High Repetition Rate High Power, Line-Narrowed Alexandrite Laser

R.C. Sam, R. Rapoport, and S. Matthews

Military Laser Products, Allied Technologies Co., 31717 La Tienda Dr., Westlake Village, CA 91362, USA

1. Introduction

This paper describes the specifications, conceptual design and development of a high repetition rate, high average power, line-narrowed, tunable alexandrite laser system. A brief description of the development of support hardware, experimental data obtained, difficulties encountered, and design modifications to resolve these problems is given. Also included is an update on current system performance.

2. Laser Design

High energy output, broadband tunability, and line-narrowed operation of alexandrite lasers have been demonstrated on separate occasions[1]. The requirements of this system bring all of these characteristics together as noted in Table 1. The significant issues to be faced were:

1. operation at repetition rates $\geq$100 Hz;
2. simultaneous attainment of spectral, spatial, and temporal requirements; and
3. operation at the long wavelength edge of alexandrite's tuning range.

Table 1 - DoE High Repetition Rate Laser Specifications

Output Energy per Pulse	600 mJ
Pulse Repetition Rate	250/n Hz (n=1 to 2,000)
Temporal Pulsewidth (FWHM)	300 ns $\leqslant \tau \leqslant$ 1 μsec
Wavelength Tuning Range	790-793 nm
Laser Output Linewidth	0.005 nm $\pm$ 0.001 nm
Maximum Allowable Out-of-Band Energy	60 mJ
Laser Output Wavelength Drift	$\pm$ 0.002 nm/hr.
Beam Profile	smooth supergaussian
Full Width Beam Divergence	10 mrad-mm

The major constraints on the design were:

1. the average power loading capabilities of Xe flashlamps;

2. thermally induced lensing and astigmatism; and

3. laser induced damage limitations.

To achieve the required specifications, a master oscillator (MO) was chosen to frequency control an injection locked oscillator (ILO). This approach was chosen in order to minimize the losses in the amplifier by removing the bulk of the frequency selective elements and to supply sufficient energy to frequency lock additional ILO's.

This approach was developed due to the large experimentally observed power gains (10^9 at 755 decreasing to 10^6 at 790 nm)[2]. These large locking efficiencies are due to the very long buildup time of alexandrite and minimal energy needed to overcome spontaneous emission in the solid angle along the laser axis.

Due to the long buildup time of alexandrite ($\approx$ 800 ns), rod quality, resonator length, and type of resonator, the ILO configuration alone controlled the spatial and temporal characteristics of the output.

In order to reach these power gains, it was necessary to suppress center band fluorescence. This was accomplished by using a 3 plate birefrigent filter. Also required was temporal stability. This requirement imposed an approximate 50 mJ output from the MO, a minimum level that insured temporal stability yet was several orders of magnitude over what was required to frequency lock the ILO.

2.1 MO Design

The MO consisted of a single 5 mm diameter by 10 cm long laser rod, a 5 plate birefringent filter with a quartz plate thickness ratio of 1:2:2:10:10 (a 1 plate is 0.51 mm thick) and a single 1 mm 25% reflective etalon. The line-narrowing technique was previously demonstrated in a laser developed for LIDAR applications[3]. The input requirement of 40 joules/pulse at 250 Hz set the output coupler reflectivity at 70% R to extract 50 mJ. Due to the long buildup time and desired 300 ns output pulse, an acousto-optic Q-switch was chosen. Spherical thermal lensing was corrected by grinding -20 cm radius of curvatures on the ends of the laser rod.

ILO Design

The design analysis of the ILO followed similar lines as those of the MO. The ILO was charged with delivering 600 mJ at a rate of 250 Hz. The low gain and high storage capabilities of alexandrite prompted the use of multiple rods in a single resonator. The design was optimized by trading off the rod temperature, rod number, and output coupler reflectivity.

The limitations to design flexibility were the flashlamp wall loading limit, set at 350 W cm^{-2} and an intracavity peak power density, of 40 MW cm^{-2}. These two constraints set limits on the rod diameter and output coupler reflectivity. In applying the power density limit, a 50% filling

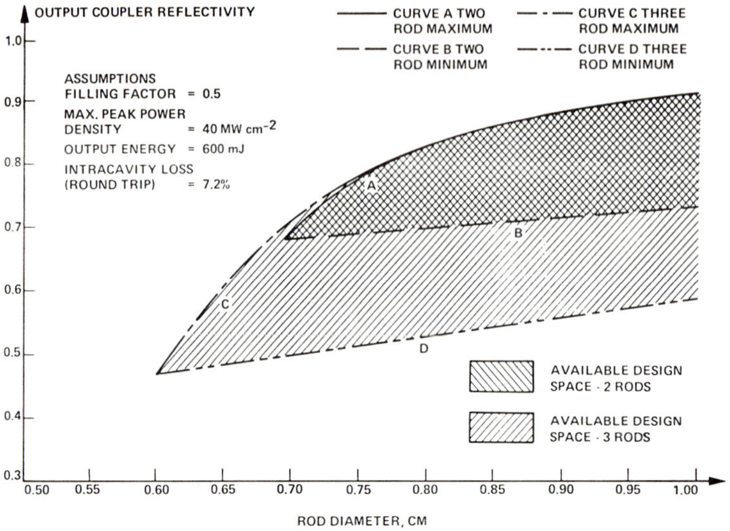

Fig. 1 Laser rod diameter versus reflectivity trade-off diagram

factor was assumed. Figure 1 shows a trade-off diagram of output mirror reflectivity versus rod diameter for two and three rods per resonator. The top curve represents the maximum and the bottom curve the minimum reflectivities for each rod diameter. The maximum curve is determined by the intracavity power density limit and the minimum by the flashlamp loading limit. The areas enclosed by the curves are the operational domain. From Fig. 1, it is apparent that the 2 rod configuration is riskier, as one is forced to use larger rod sizes which are not readily available, and higher reflectivities. The 3 rod configuration allows the use of 6.35 mm rods, which are available.

The gain of alexandrite is strongly influenced by temperature. Fig. 2 shows the temperature dependence of the emission and excited state absorption cross sections at 790 nm. The ground state absorption cross section, σ_a, is two orders of magnitude smaller than σ_a and σ_e. In choosing an optimal temperature, a reduction in fluorescent decay time with temperature rise was considered. The improvement in efficiency due to higher gain must

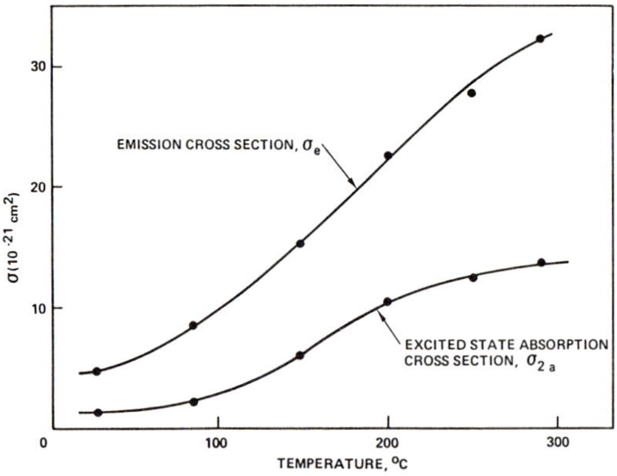

Fig. 2 Temperature dependence of the emission & excited state absorption cross sections at 790 nm.

be balanced against the efficiency loss from fluorescent decay as the flashlamp pulse duration is limited by lifetime and plasma temperature constraints.

A trade-off study was performed to evaluate the ILO performance at different temperatures. Using 6.35 mm rods and matching flashlamps with 31 J input energy, the attainable output energy at these temperatures was plotted versus the number of stages (rods). The analysis indicated that a 90°C operating temperature would offer the best performance. This temperature is also attainable with water cooling.

An energy balance diagram for a 3 rod ILO is shown in Fig. 3. The efficiency numbers used in this diagram are estimates from the literature or calculated. As will be shown later, these efficiencies are in agreement with the experiment. Using a 54% reflecting output coupler, this model projects a threshold energy at 790 nm of 125 J and slope efficiency of 2.3%.

The baseline design optical layout is shown in Fig. 4. It consists of 3 parts: the MO, ILO, and intermediate optics. The intermediate optics were designed to inject the MO output through the rear mirror of the ILO. This mirror transmits 2% of the MO pulse. To prevent back reflection of

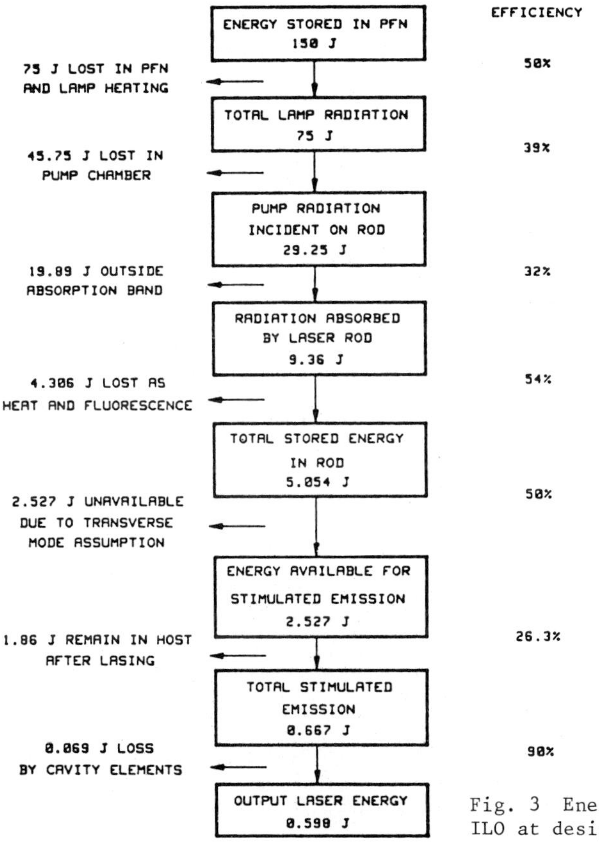

Fig. 3 Energy balance diagram for ILO at design point.

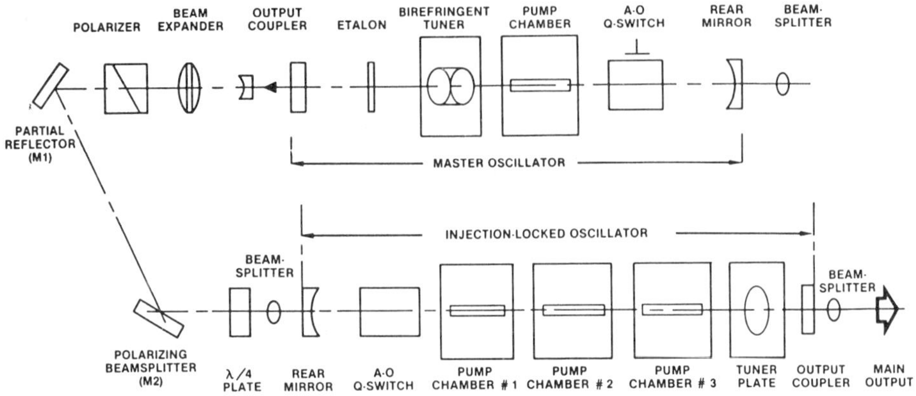

Fig. 4 Design point laser optical layout diagram.

the MO output into itself, polarizers and wave retardation plates were used. The beam splitters indicated are for beam sampling for diagnostic purposes.

3. Laser Development and System Considerations

In parallel with the laser development, support subsystems were developed. These included the laser table, power conditioners, and cooling units. The laser is required to operate in a plantlike environment, which imposes two additional requirements. The laser must be sealed off from the surrounding environment during operation as cleanliness is necessary for

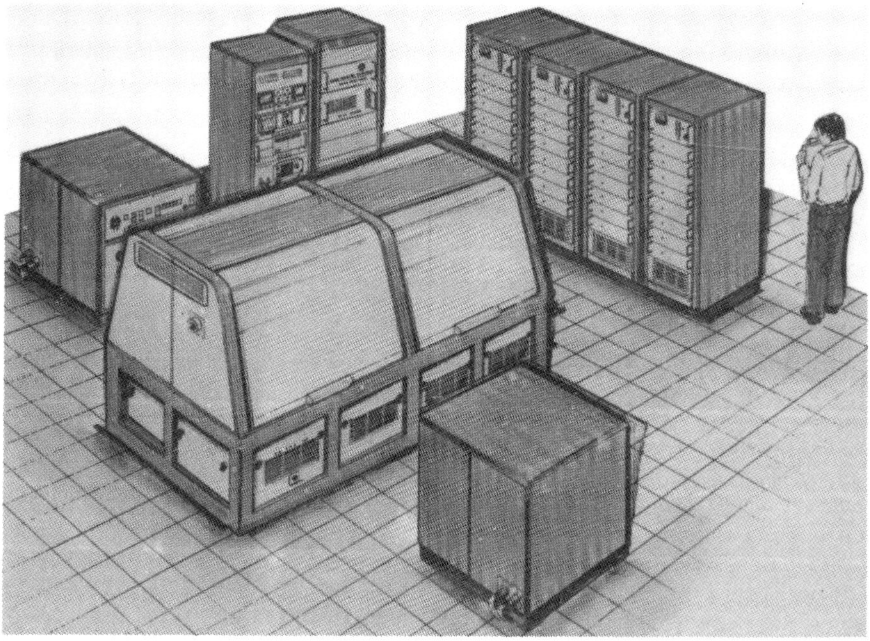

Fig. 5 Laser system conceptual drawing.

reliable operation, therefore all operational controls, such as optics fine alignment and tuning, must be performed remotely (an enclosed MOPA laser structure, and a centralized system control were developed). Secondly, many of the laser characteristics must be monitored in real time, so sensors were located on a separate diagnostics table directly below the laser table. A conceptual drawing of the finished system is shown in Fig. 5.

The following addresses the major technical issues encountered during the course of development. Some problems were solved by improvements in design, and some by a change of approach.

3.1 Flashlamps - The requirement of this system demanded that the flashlamps used be capable of operating under a high thermal load and exhibit long life. To illustrate, an hour of operation at 250 Hz is equivalent to nearly 10^6 shots. While the operating point is close to the thermal fracture limit, it is way below the explosion limit. The design of the flashlamp uses a thin (0.5 mm) wall for thermal stress reduction and long tungsten electrodes for more efficient heat removal. Laboratory tests operating the flashlamp at design point were performed. Many lamps exhibited lifetimes in excess of 5×10^7 shots [4] (lifetime is defined as the time it takes for the light output to be reduced to 50% of those of a new lamp). Actual laser tests using these lamps show similar results.

3.2 Pump Chamber and Reflector Development - From a technical standpoint, one of the most critical tasks in the program has been the design of an efficient pump head which satisfies the need for good maintainability and reliability. The most difficult problem encountered in this effort has been providing sufficient heat removal for the flashlamps, cavity reflector, and laser rod, while at the same time maintaining the rod at an effective operating temperature of 90°C. In addition, an efficient pump reflector had to be developed which would not degrade under the extremely high heat loads in the laser head.

The pump cavity reflector issues were solved by developing a monolithic double ellipse pyrex reflector. The coating is applied to the outside of the pyrex sleeve, and consists of successive layers of dielectric binder, silver, copper, chrome, and a final protective layer of SiO_2. With proper handling during the assembly process, it has been operated in excess of 2×10^8 shots at a power loading of between 10 and 15 kW without any noticeable deterioration in performance. These reflectors have been successful due to the excellent thermal transfer of the thin metallic layers.

The pump chamber assembly had to meet several requirements. They were: water-cool the laser rod at 90°C and the flashlamps at 25°C; provide fast flow velocities within the flow tubes; cool both surfaces of the pyrex, and water-cool the flashlamp electrodes. This required deionized water since the lamp electrodes were submerged and also an 8 gal/min flow rate within the tubes.

A "shower head" design distribution plate was incorporated into the pump chamber to evenly distribute the cooling water to the various surfaces. This is a flow plate perforated by many small holes directing the flow parallel to the lamps and rod.

3.3 Thermally-Induced Lensing and Astigmatism - Like all liquid cooled solid state laser materials, alexandrite suffers from thermal gradient lensing when pumped. At relatively low power input, the lensing rate is on the order of 0.5 diopters per kW. The thermal lensing induced by high power pumping can be effectively corrected by polishing concave curvatures on the laser rod ends, such that resulting geometrical lensing counterbalances the thermal gradient lensing.

Fig. 6 shows the rod end radii of curvature versus input power by equating the negative geometric lensing to the positive thermal lensing. The validity of the curve was tested by measuring the input power at which rods with given radius of curvature start to lase. Instead of a point, a line indicates a range of turn-on power for several rods with the same curvature. It is found that, while the agreement is good for rods with relatively long radii of curvature, for short radii the turn-on power is smaller than predicted.

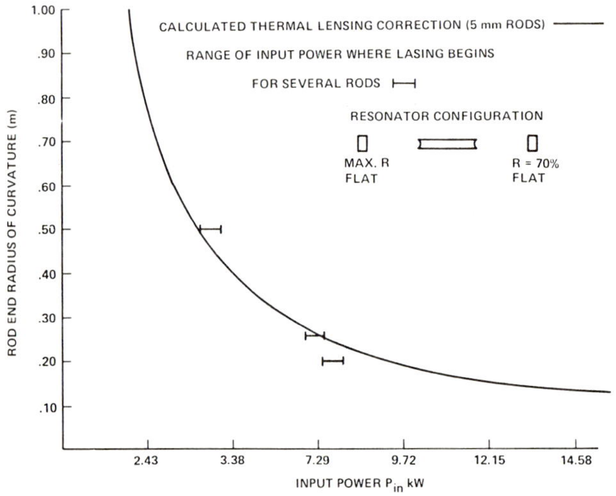

Fig. 6 Zero lensing correction radii of curvature as a function of input power.

To investigate the observed discrepancy, thermal lensing data of curvatured rods at powers above the compensation points were measured. The results for 6.35 mm rods are shown in Fig. 7. The lensing rate appeared to increase monotonically with input power, which qualitatively agrees with the lensing correction discrepancy observed.

Another observation of curvatured rods under heavy pumping without additional compensation, is that the filling factor is less than $<$50%. This anomaly is caused by thermally-induced astigmatism. An active measurement of astigmatism was performed by placing a laser rod inside a pump chamber in one arm of a Mach-Zender interferometer.

The interference pattern was recorded with a video camera. With a curved end, c-axis rod, the interference pattern changes from circular to elliptical, then to straight fringes along the a-axis. A saddle-shaped fringe pattern subsequently forms and then straight fringes appear along the b-axis. Thermal lensing in alexandrite thus exhibits a strong astigmatism, the positive lensing being stronger in the a-axis plane than that

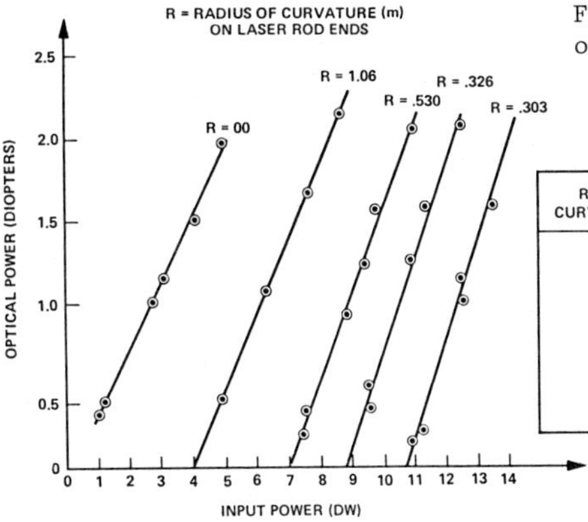

RADIUS OF CURVATURE (m)	LENSING RATE (DIOPTERS KW^{-1})
∞	0.42
1.06	0.46
0.53	0.51
0.326	0.54
0.303	0.61

Fig. 7 Thermal lensing rate dependence on input power.

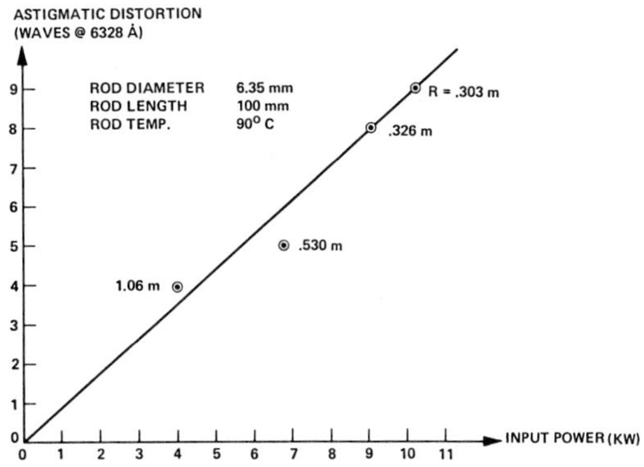

Fig. 8 Astigmatic distortion measurement versus input power.

in the b-axis plane. This asymmetry is not due to the rod orientation relative to the flashlamps, as the same fringe pattern rotates with the rod as the rod is rotated relative to the lamp plane. By taking the peak-to-valley fringe count at the saddle point, a quantitative measure of the thermally induced astigmatism is obtained. This is shown in Fig 8.

Obviously the observed astigmatism is the main culprit for a small fill factor. It also renders the 3 rod ILO scheme very difficult to implement under high power pumping. A two rod resonator is easier to compensate. The two schemes being considered are: a) passive compensation using intracavity cylindrical lenses and, b) active compensation with the b-axis of one rod rotated 90° relative to the second rod with insertion of a polarization rotator between the rods to recover the lasing polarization.

3.4 Laser Induced Damage - As in all solid state lasers, a fundamental limitation to alexandrite output power in the Q-switched mode is laser-induced damage. The damage site usually starts out as a small pit on an AR coated rod end at the most intense spot of the beam which quickly expands to a crater of sizable dimensions when the laser is operated under repetition rate conditions. At high (>100 pps) repetition rates, laser damage rapidly leads to self-destruction of the laser rod.

In laboratory tests, we found that the average lifetime of a damage-free laser rod at peak intensity in excess of 200 MW cm^{-2}, to be on the order of a few thousand shots; those at 100 MW cm^{-2} last 10^5 shots; and those below 50 MW cm^{-2} to live almost indefinitely. These numbers, though crude, are useful in guiding design engineering.

Several attempts were made to reduce hot spots in the beam and to increase the filling factor. The first attempt replaced the rear mirror with a porro prism. This helped average the high and low gain areas, broke up the lowest order laser modes and provided alignment stability. The second attempt was a "dot" mirror. This is akin to a simplistic apodized mirror having a low reflectivity in the center few millimeters and a higher reflectivity on the outer edges in order to encourage lasing of higher order modes.

4. Program Status

The MO development has achieved all the spectral control and output energy requirements using the proposed techniques. However, due to thermal gradient lensing and accompanying astigmatism, the highest repetition rate attained was 200 Hz. Above this rate, the filling factor was reduced such that the intracavity power density was in excess of 100 MW cm^{-2} when operated at 50 mJ output energy.

Figure 9 shows a plot of the output coupler reflectivity versus threshold input energy of the ILO at 790 nm and 90°C. The data from 1, 2, and 3 rods appear to fall on the same line indicating that the absorption and scattering losses from individual rods are negligible, which agrees with the measured intracavity loss of 7%. The threshold for a 54% output reflector is 124 J, in good agreement with the design point model.

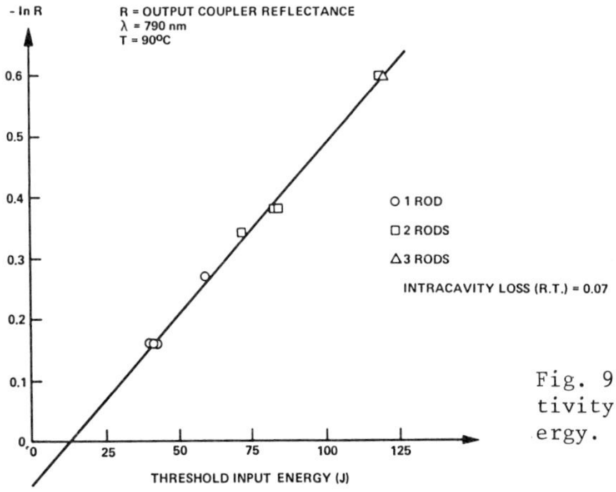

Fig. 9 ILO output coupler reflectivity versus threshold output energy.

As discussed in Section 3.3, the 3 rod ILO design encountered significant difficulties under high input power loading as design point slope efficiencies cannot be realized due to increasingly small filling factors. At low repetition rates (10-20 Hz), slope efficiencies of 1% have been observed. It appeared that the long resonator length had a tendency to suppress high order spatial modes.

4.1 Current Results

Currently, an alternative configuration is being investigated. It consists of a pair of tandem rod, line-narrowed resonators each operating at 125 Hz. The output from the two lasers will be temporally interlaced and spatially multiplexed via a polarization beam combiner. Operating with flat end rods at 10 Hz demonstrated that the linewidth and energy requirements were met.

Two major breakthroughs have paced the experimental progress. The first was a significant increase in the filling factor by employing cylindrical lens correction at a rate of 0.1 diopters/kW. This has increased the filling factor close to 50% at the full operating input. The second advancement was a novel Q-switch technique in which the output was extracted in several closely-spaced pulses instead of a single pulse. This technique permits the extraction of three Q-switch pulses in a 3-5 μsec time window. Since the energy is emitted in several pulses instead of one, the individual pulses are temporally wider, thus the result is a significantly reduced peak power to about 30 MW cm^{-2}.

Since the pulses are longer in duration due to larger buildup times, the laser linewidth is reduced, closely resembling that of a long pulse device. Details of this Q-switching technique will be published in a separate article.

Using both techniques in a 2 rod cavity, the laser has now made several 30 minute runs at 400 mJ 125 Hz @ 790 nm without line-narrowing etalons. Line-narrowed operation has been achieved at 124 Hz, 300 mJ at .004 nm also for several 30 minute runs.

Acknowledgements

The laser system described was developed for the Department of Energy under contract no. 9X43-H49309. Their permission to publish this material is appreciated. We would like to thank J. J. Yeh, Ralph Smith, James Dobbins, Murray Finegold, and Louise Tanguy for their contributions.

References

1. J. C. Walling, O. G. Peterson, H. P. Jenssen, R. C. Morris, and E. W. O'Dell, IEEE Journal of Quantum Electronics, QE-16, 1302 (1980).

2. W. R. Rapoport, J. J. Yeh, and R. C. Sam, Paper ThR4, presented at CLEO '83, Baltimore, MD., 1983

3. R. C. Sam and F. P. Roullard III, EOSD Magazine, pp. 37-40, July 1982.

4. Report R-ILC-83-10, by ILC Technology, August 1983.

Alexandrite Laser Systems: Scaling to Kilowatt Powers

David C. Johannsen
575 Hampshire Road, 221, Westlake Village, CA 91361, USA

With alexandrite laser systems producing tens of watts now developed, it is interesting to consider how powerful a system could be developed through the scaling up of existing designs. The most logical starting point for such a scaling is the 150 W system,which Allied is developing under contract to the Department of Energy (DOE), which is currently the most powerful alexandrite system in preparation. In this paper I discuss how an optimization of parameters for maximum power production coupled with larger alexandrite elements might allow scaling to kilowatt power.

The DOE baseline design [1] is a high rep rate system producing narrow linewidth pulses of 300 ns duration, tunable from 790 to 793 nm. It is composed of two oscillators, each employing two 0.635 x 10 cm rods to produce 600 mJ per pulse at 125 Hz, for a total power per oscillator of 75 W. The two oscillators are then temporally encoded to produce 150 W at 250 Hz. Figure 1 shows schematically one of the oscillators. This two rod oscillator design is currently being tested as a means of ameliorating thermal lensing problems encountered with the original three rod oscillator baseline design [1].

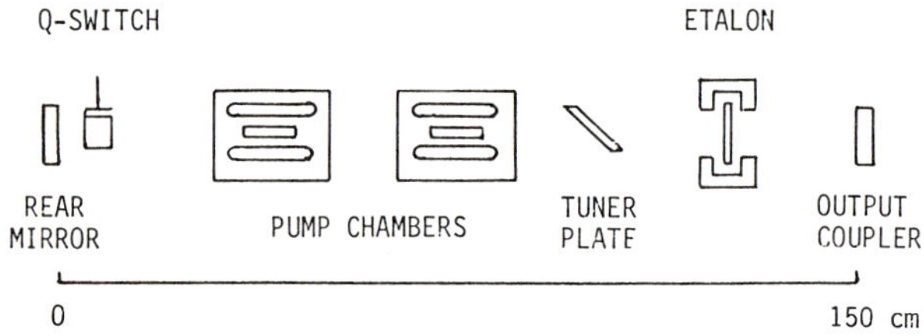

Fig. 1. Schematic diagram of the DOE oscillator

Work to date on this system [1] has shown that the power output corresponds to measured alexandrite cross sections [2] and that the power limits are driven by damage thresholds, provided that the total power input to the laser head does not exceed the limit where thermal lensing predictions break down [1]. This breakdown point appears to be somewhat under 10 kW per head for the DOE design; the important quantity is actually the total power absorbed by the rods, which depends on more than just the power input to the flashlamps.

The scaling I will discuss is based on a system which retains the 300 ns pulsewidth and two alexandrite element design of the DOE system. For simplicity I have chosen a rep rate of 100 Hz. Peak power has been limited to 150 MW/cm^2 from damage consideration, and the observed breakdown point for thermal lensing predictions has been retained for 0.635 cm rods; for other rod diameters I have scaled the power into the rod as the square of the rod diameter, as it is the surface area of the rod which determines the cooling ability of the system, and this scales as the square of the rod diameter.

Increased power can be realized by optimizing the operating wavelength of the laser system--the DOE system wavelength of 790 to 793 nm was not chosen to maximize power. It is also important to optimize with respect to temperature. Further gains may be realized by using larger rods, or employing slabs; slabs allow a larger material cross-section to be utilized, and should ameliorate thermal lensing problems. The actual calculations were done numerically, with the appropriate laser rate equations [1] augmented with enforced limits on peak power to prevent damage thresholds from being exceeded, etc.

Figure 2 shows how the output of a DOE-like system scales the wavelength. The basic layout of the DOE baseline has been retained, but for each wavelength the length of the resonator cavity and the reflectivity of the output coupler has been optimized to maximize pulse energy. It may be seen that the highest energy output is in the 760 to 770 nm region. In this region the primary limit on performance is the damage threshold--allowing a higher peak power would substantially increase the system power, but at the cost of significantly reduced lifetime. In what follows I adopt 765 nm as the optimum wavelength, although the value of the optimum wavelength varies with temperature.

Figure 3 shows how performance at 765 nm varies with rod temperature. Again, resonator length and output coupler reflectivity have been optimized for each point. It is seen that performance peaks at about 90°C. This peak is due to limitations on available flashlamps. The gain in alexandrite actually continues to increase with temperature [2], but at the same time the fluorescence lifetime decreases. As there is a minimum pulsewidth for which the flashlamps can be operated efficiently and reliably, the fluorescence losses increase with temperature, eventually offsetting the increased gain. I have adopted a nominal value of 70 μs as a minimum pulse duration for the flashlamps.

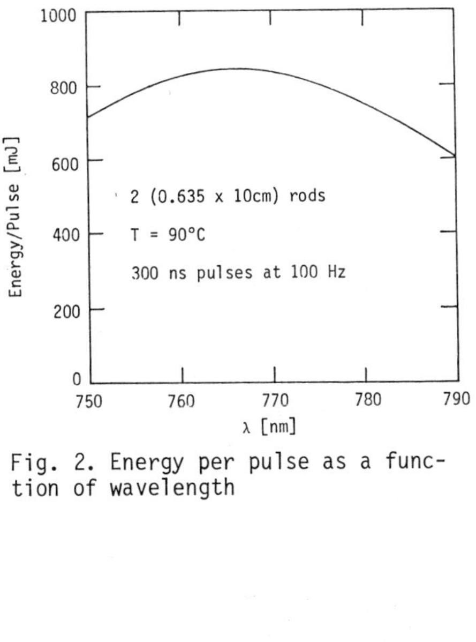

Fig. 2. Energy per pulse as a function of wavelength

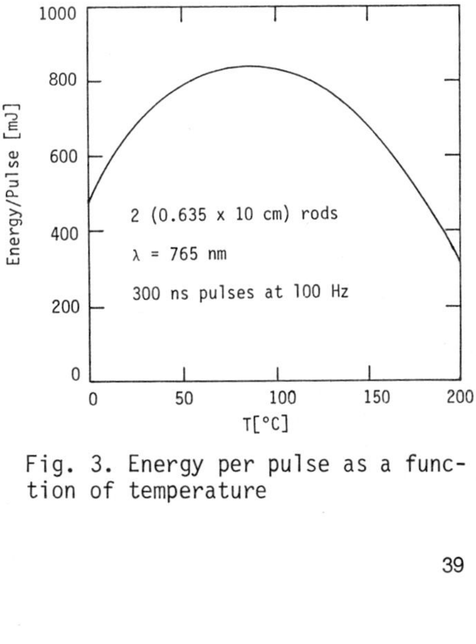

Fig. 3. Energy per pulse as a function of temperature

Figure 4 shows the increase in power as the rod diameter is increased. The Cr^{3+} concentration is adjusted to insure uniform pumping for all diameters. Figure 5 then shows how the power per unit lasing area approaches 1 kW/cm^2 for 1.1 cm rods. The largest rods fabricated to date have a diameter of 1 cm. (The lasing area does not include a 0.9 mm annulus at the edge of the rod, which has been observed not to lase in alexandrite). The power extracted per unit area appears to continue to increase as rod diameter increases, but at some point the limiting factor ceases to be the damage threshold of the material and becomes the energy storage limit, which actually decreases with diameter because the Cr^{3+} concentration must be decreased to insure uniform pumping. Therefore, 1 kW/cm^2 seems a good value to adopt for the power which can be produced by a large cross-section (i.e., slab geometry) system.

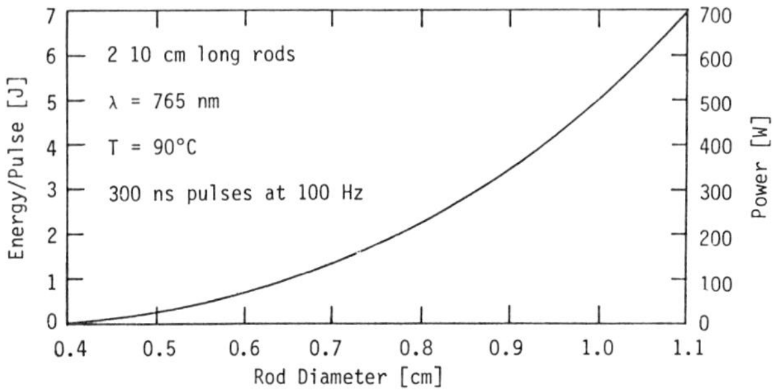

Fig. 4. Energy per pulse as a function of rod diameter

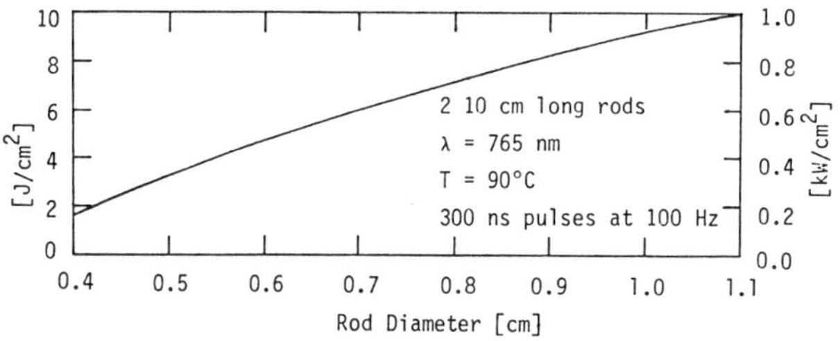

Fig. 5. Energy per pulse and power per unit lasing area as a function of rod diameter

Considering the above, we see that using 1 cm rods in a DOE-like systems implies that two systems will produce on the order of 1 kW. By employing slabs, however, substantially higher power can be obtained due to the larger material cross-section. A system with slabs the size of an already fabricated 6 x 12 x 96 mm alexandrite slab would yield about 700 W. Larger slabs would, of course yield even more power--for example, a 1 x 2 cm slab would have 2 cm^2 of lasing area, implying 2 kW, and a 2 x 5 cm slab would have 10 cm^2 of lasing area, implying 10 kW.

In summary, we have seen that relatively modest scaling of current technology may make possible alexandrite laser systems with powers between one and ten kilowatts, which implies that the clumping of multiple oscillators could produce a power on the order of 100 kW, if needed. This scaling assumes that no further difficulties with thermal lensing are encountered in the two rod oscillator design, and that no unexpected effects are found to be associated with larger alexandrite elements.

References

1. R. Sam, R. Rapoport and S. Matthews, elsewhere in this volume.

2. M. L. Shand and J. C. Walling, IEEE J. Quantum Electron. QE-18, 1152 (1982).

An Overview of NASA Requirements for Tunable Solid State Laser Systems and Technology

F. Allario and B.A. Conway

NASA Langley Research Center, Hampton, VA 23665, USA

A broad overview of NASA's science and application needs for Lidar Systems Technology from space, airborne, and ground-based platforms will be presented. In particular, future needs for Lidar Technology in conducting scientific investigations in the NASA's Earth Sciences and Applications Program from space platforms will be highlighted. Examples of scientific investigations utilizing Lidar systems within NASA will be presented for measuring atmospheric aerosols, meteorological variables and atmospheric species. In this paper, scientific and instrumentation results will be used to project a shopping list of instrumentation requirements for "All" Solid State Lidar Systems for long-duration missions from space platforms and high flying aircraft. Some current thoughts on future research directions in developing "All" Solid State Lidar Systems for NASA's science investigations will be presented, with particular emphasis on vibronic lasers.

1. Introduction

The purpose of this paper is to provide an overview of emerging needs and requirements for tunable solid state lidar systems technology to meet NASA science, applications, and technology programs. In addition, the current status and plans for technology thrusts in solid state laser technology will be discussed, with emphasis on vibronic lasers. The format to be used for this presentation is as follows. In the next section, (Section 2), a perspective of atmospheric science and applications needs from Space Station observing platforms will be presented. This will be followed by section 3, highlighting current plans and status for conducting a series of scientific investigations of the earth's atmosphere utilizing a lidar system on a NASA ER-2 aircraft. Section 4 will include some comments on the current thrusts being planned for a technology program in solid state lidar to support the ER-2 investigations and to support a Space Station Earth Observing System (EOS).

2. Science and Application Needs

With the advent of the Space Station Program within the National Aeronautics and Space Administration (NASA)[1], significant opportunities to conduct lidar experiments from space should emerge to investigate geophysical properties of the earth, including land features, ocean properties and the atmosphere. Since land, oceans, and the atmosphere are coupled, many of the foremost earth sciences questions can be addressed in isolation (i.e. piecemeal), but ultimately require observations of the system as a whole. The interdisciplinary questions require observations across the broad spectrum of earth sciences. An important example is the study of biogeochemical cycles, the processes by which key chemical elements are transformed and exchanged among the soils, biota, inland waters, oceans, and atmosphere. A second example is the hydrologic cycle, in which water is cycled among the oceans, atmosphere, cryosphere, and land surfaces while undergoing phase changes and redistributing energy within the earth system. While these are distinct problems, they are also important parts of the study of climate in which the interactions among atmospheric composition, soil moisture, surface albedo, cloud amount, ocean heat transport, and atmospheric dynamics couple together to determine the environment in which life takes place. As part of the Space Station concept, a study is currently in progress on the use of a Space Station-supported platform in polar orbit to address multidisciplinary needs of earth sciences, and which is to include a variety of remote sensor instruments covering the electromagnetic spectrum from the ultraviolet through the microwave range. Several lidar investigations have been identified as potential candidates to measure atmospheric winds and improve scientific understanding of atmospheric chemistry, dynamics and meteorology. Figure 1 is a list of candidate instruments that have been identified as needed to meet scientific requirements of an Earth Observing System (EOS) in a polar orbit. These projected needs were developed by a multidisciplinary scientific working group convened under NASA sponsorship.[2] Along with the preliminary listing of candidate instruments, Figure 1 lists measurement needs generated by the working group, with gross estimates of weight, size, power and data requirements for each instrument. The fourth column indicates the heritage within the NASA Science and Applications Program. In particular, many of the candidate instruments that have been scoped have a strong heritage on satellite missions, or as Shuttle payloads. In the case of lidar, extensive aircraft investigations have been flown to obtain range-resolved atmospheric profiles of aerosols, ozone and water vapor.[3,4] In addition,

INSTRUMENT	MEASUREMENT	ESTIMATES PARAMETERS				HERITAGE
		WEIGHT (Kg)	SIZE (m^3)	POWER (Watts)	DATA (MBPS)	
1. MODERATE RESOLUTION IMAGING SPECTROMETER (2.1.1)	1Km RES:BIOLOGICAL CYCLES	300	2.5	350	<5	CZCS/AVHRR/STUDY
2. HIGH RESOLUTION IMAGING SPECTROMETER (2.2.2)	30n RES:BIOLOGY/GEOLOGY	400	4.5	400	300	SISEX/TM/MS/STUDY
3. HIGH RESOLUTION MULTIFREQUENCY MICROWAVE RADIOMETER (2.1.3)	HYDROLOGY; SOIL MOISTURE	220	---	235	.05	LAMMR/STUDY/SMMR
4. LIDAR ATMOSPHERIC SOUNDER AND ALTIMETER (2.1.4)	ATMOSPHERIC CORRECTION	600	18n3	2000	.005	A/C-STUDY/BALLOON
5. SYNTHETIC APERTURE RADAR (2.2.1)	LAND, ICE, VEGETATION	400	0.4	4000	300	SEASAT/SIR
6. RADAR ALTIMETER (2.2.2)	TOPOGRAPHY	150	0.1	200	.01	SEASAT/TOPEX
7. SCATTEROMETER (2.2.3)	WIND STRESS	150	0.2	200	.01	SEASAT/NROSS
8. CORRELATION RADIOMETER (2.3.1)	SPECIES:NH_3, NO, OH, O_3	100	0.5	100	<.01	MAPS/DCR
9. INTERFEROMETER/SPECTROMETER (2.3.2)	NO_3, CH_4, H_2S, ETC.	100	<0.1	<100	<.03	IRIS
10. DIFFERENTIAL ABSORPTION LIDAR (2.3.3)	SPECIES					A/C-/STUDIES
11. DOPPLER LIDAR (2.3.4)	TROPOSPHERIC WINDS					STUDIES
12. IR RADIOMETER (2.3.5)	SPECIES;LIMB;O_3,N_2O, ETC.	250	0.1	100	<.02	LIMS/LRIR
13. INTERFEROMETER SPECTROMETER (2.3.6)	SPECIES;LIMB;$CFCL_3$, ETC.	440	3.6	<100	<.01	CLAES
14. IMAGING SPECTROMETER (2.3.7)	SPICIES; LIMB					SIS/STUDIES
15. SUBMILLIMETER SPECTROMETER (2.3.8)	SPECIES; LIMB;CH,HCL,HF					NEW TECHNOLOGY
16. MICROWAVE LIMB SCANNER (2.3.9)	SPECIES; LIMB; CLO, O_3	200	5.4	150	<.01	MLS/UARS
17. FABRY PEROT INTERFEROMETER (2.3.10)	WIND & TEMPERATURE	<100	0.15	<100	<.01	HRDI
18. MICHELSON INTERFEROMETER (2.3.11)	WIND & TEMPERATURE	<100	0.3	<100	<.01	TWM/CLIR
19. SOLAR MONITOR (2.4.1)	SPECTRUM	<100	<0.1	<100	<.001	SUSIM
20. SOLAR CONSTANT MONITOR (2.4.2)	OUTPUT	<100	<0.1	<100	<.001	ACR
21. MAGNETOSPHERE PARTICLES (2.4.3)	PARTICLES	<100	<0.1	<100	<.01	PROBES
22. MAGNETOSPHERE CURRENTS/FIELDS (2.4.4)	CURRENTS & FIELDS	<100	<0.1	<100	<.01	PROBES
23. LIGHTNING DETECTOR (2.4.5)	LIGHTNING PARAMETERS					DMSP
24. AUTOMATED DATA COLLECTION & LOCATION (2.4.6) SYSTEM	MISCELLANEOUS FROM BOUYS	<300	<0.1	<200	<.02	ARGOS

Fig. 1. Preliminary instrument types candidate list - EOS

an extensive scientific study on lidar was conducted by NASA to scope a series of scientific investigations that could be conducted from Space Shuttle[5], and a parallel engineering study was conducted to identify systems level requirements and technology readiness of lidar technology for an Atmospheric Shuttle Lidar Facility.[6] However, to date no lidar experiment has been conducted from space to study atmospheric properties.

In general, the needs for atmospheric measurements identified in Figure 1 include the troposphere and stratosphere.[7 8] However, as new aerospace vehicles are defined for the future, the properties of the atmosphere from 70 to 100 km need to be further understood; in particular, measurements of molecular density are desirable in this region to predict drag properties of advanced aerospace vehicles. Ground density lidar measurements which probe this portion of the atmosphere have been conducted.[9] There may be a future need to conduct experiments from space to obtain an improved capability to measure molecular density from 70 to 100 km, or to provide a "look ahead" predictive capability from a spaceborne lidar system to identify optimal conditions for entry into the earth's atmosphere.

Another promising area for the use of solid state laser technology is: rendezvous and operations in proximity of the central Space Station, where sensors are required for tracking and location, rendezvous, docking and berthing of satellites and orbiting maneuvering vehicles, as well as orbital transfer vehicles. An important technology to provide sensor information for range and range rate (velocity) measurements is laser radar. For this application, broad wavelength tunability may not be a critical technology requirement for the laser transmitter; however, a long lifetime, reliable laser technology will be required with a high degree of compact engineering design. Technology requirements for laser radar in rendezvous and proximity operations are more closely related to requirements for laser range finders and target designators rather than lidar and will not be discussed in this presentation. A Workshop on "Rendezvous and Proximity Operations" will be held at the NASA Lyndon B. Johnson Spacecraft Center on October 16-19, 1984 with a major objective to assess technology and operational readiness of sensors to meet requirements for proximity operations. A major agenda item will be laser and millimeter radar systems and technology. Proceedings from the workshop will be available following its conclusion.

3. NASA ER-2 Science and Technology Experiments (Project "LASE")

In this section, an overview of current NASA plans to conduct a series of observations with lidar from the NASA ER-2 aircraft will be presented. This experiment which is called project LASE (Laser Atmospheric Sensing Experiment) was conceived with several objectives in mind. The primary objective is to conduct a series of scientific investigations of aerosols, atmospheric species and meteorological parameters (pressure, temperature, water vapor) at an altitude higher than conventional remote sensing aircraft. This will extend the geographical range within a mission, at an altitude well above the tropopause. This provides atmospheric scientists increased opportunities to conduct meaningful investigations of atmospheric chemistry and dynamics for global scale modeling of the atmosphere. A second objective of project LASE is to serve as a high-flying test bed to demonstrate the capability of new laser technology to conduct scientific experiments of the atmosphere utilizing the DIAL (Differential Absorption Lidar) technique.[10] For the latter objective, the instrument design is to be configured as modular and flexible at the subsystem level, in order to incorporate new laser technology as it emerges from the laboratory with minimal impact on the total lidar design.

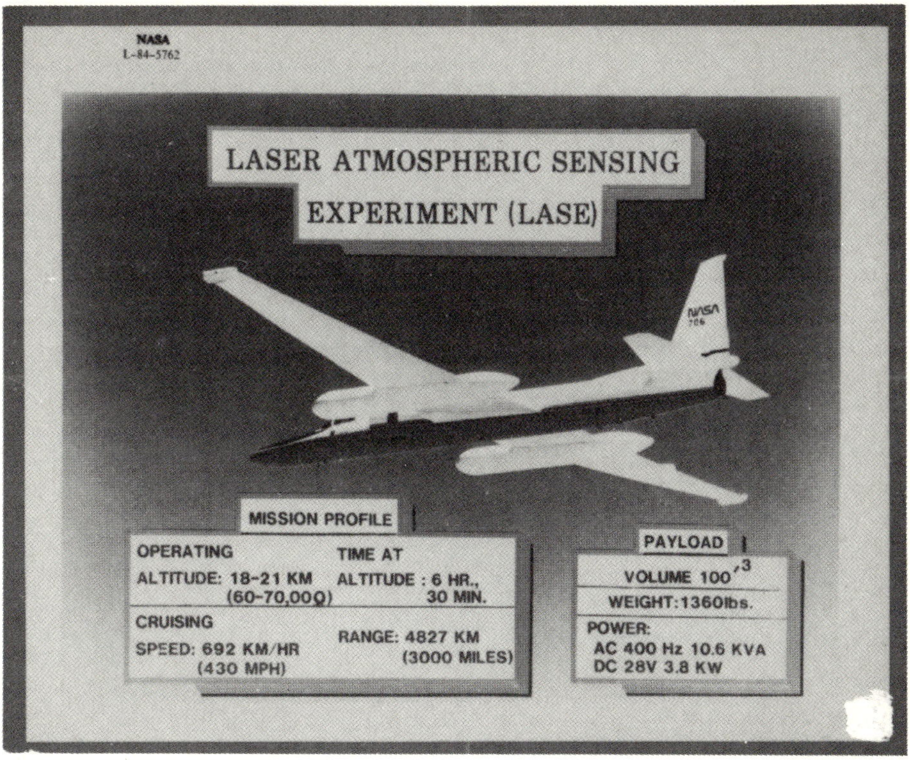

Figure 2

Significant operational characteristics of the ER-2 aircraft are shown in Figure 2. In addition to these operational characteristics, it should be noted that the ER-2 Lidar experiments must be operated in a completely autonomous mode, without any services provided by the pilot other than a "power up" and "power down" switch sequence. This high degree of autonomy raises some interesting challenges with respect to the instrument design for a variety of subsystem functions including wavelength calibration and control, detector linearity and stability, as well as thermal control of the laser environment. However, if successfully executed, development of this concept for the ER-2 aircraft will provide a significant technology and design methodology for development of future space platform and Space Shuttle lidar instruments.

In Figure 3, a conceptual drawing of a lidar system configured for the ER-2 aircraft is shown. The entire instrument package has been designed to mount into the "Q" bay of the aircraft. The telescope, laser transmitter and receiver optics and detector electronics are to be mounted on a graphite-epoxy optical bench in the fuselage of the "Q" bay. On the auxiliary deck, power, signal conditioning, and data system electronics are accomodated to support the lidar experiment and to provide an improved thermal environment for the optical subsystems.

The insert in Figure 3 lists some of the potential experiments that could be conducted from the ER-2 aircraft. These experiments were initially

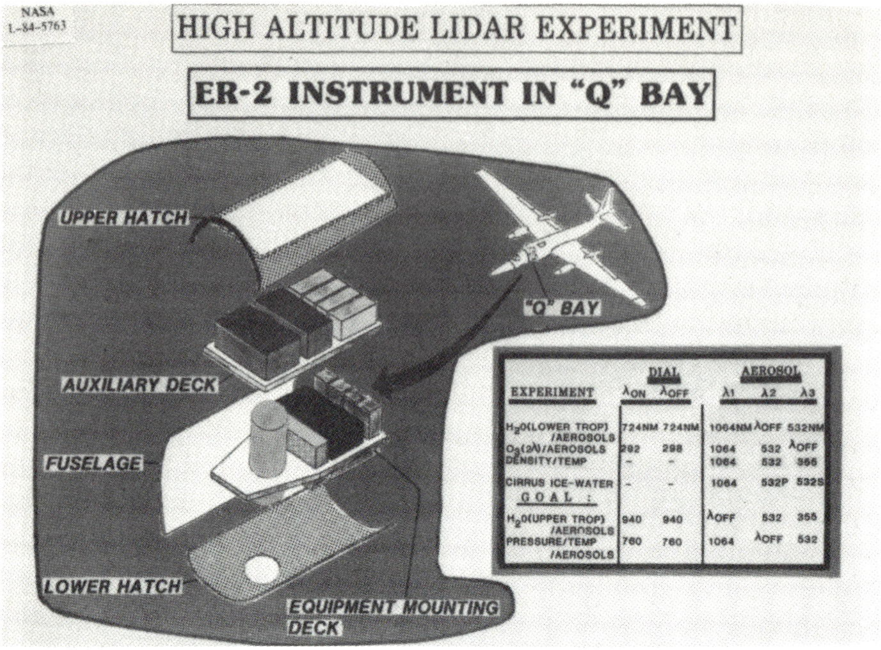

Figure 3

conceived assuming that DIAL measurements would be conducted using a double-pulsed Nd:YAG laser, frequency-doubled to 532 nm, and used to pump a dye laser. The dye laser is to provide two sequential pulses spaced 400 μs apart, at two different wavelengths; one wavelength corresponds to an absorption line of an atmospheric molecular line, and the other is spaced sufficiently far apart from the absorption line of interest and in an "atmospheric window" to provide the baseline backscatter signal from aerosols and molecules in the atmosphere. By appropriate selection of laser wavelengths absorbed by atmospheric molecules (λ_{on}), and non absorbed (λ_{off}), one could infer the total column content of an atmospheric species from the ground to the altitude of the aircraft through the utilization of the lidar equation.[10] In addition, by gating the temporal response of the lidar signal returns, a range-resolved measurement of atmospheric molecules as a function of altitude can be obtained over the mission. The DIAL technique currently has provided valuable scientific measurements of aerosols[8], ozone[4] and water vapor[11] from the NASA CV-990 and Electra aircraft. In addition, the ER-2 lidar experiment was designed to utilize a portion of the laser energy at 1064 nm and 532 nm to simultaneously measure aerosol vertical profiles at these two wavelengths.

As noted above, the insert in Figure 3 lists candidate ER-2 lidar experiments and wavelengths. Several important wavelengths are identified including: 721 nm for tropospheric water vapor, 940 nm for upper tropospheric water vapor, 760 nm for pressure/temperature, and 292 and 298 nm for ozone. Again, this series of experiment scenarios are conceived on the premise of a Nd:YAG laser, doubled in frequency to 532 nm, and used to pump a dye laser. The total lidar instrument is a four channel instrument with capability to conduct only one of these experiments on an individual mission.

TABLE 1. ESTIMATE OF ER-2 LIDAR SYSTEM PARAMETERS

ESTIMATE OF ER-2 LIDAR SYSTEM PARAMETERS		
Experiment	Transmitter	Receiver
GENERAL	LASER REPETITION RATE: 5 Hz LASER PULSE LENGTH: 15 ns	TELESCOPE DIAMETER: 35 cm TELESCOPE COLLECTING AREA: 0.086 cm^2 FIELD OF VIEW: 1 mrad
WATER VAPOR DIAL:	ENERGY:* 54 mJ (721 nm) OR 19 mJ (940 nm) LINEWIDTH: < 1.5 pm	OPTICAL EFFICIENCY: 10^{-2} (721 nm) OR 4×10^{-3} (940 nm) SPECTRAL BANDWIDTH: 0.5 nm
OZONE DIAL:	ENERGY:* 30 mJ (300 nm) LINEWIDTH: < 4 pm	OPTICAL EFFICIENCY: 8×10^{-2} (300 nm) SPECTRAL BANDWIDTH: 10 nm
AEROSOL: (SIMULTANEOUS WITH H_2O AND O_3 DIAL)	ENERGY: 300 mJ (1064 nm) LINEWIDTH: < 1 pm	OPTICAL EFFICIENCY: 4×10^{-3} (1064 nm) SPECTRAL BANDWIDTH: 0.1 nm
DENSITY/ TEMPERATURE:	ENERGY: 300 mJ (1064 nm) 300 mJ (532 nm) 60 mJ (355 nm) LINEWIDTH: < 1 pm *EACH DIAL WAVELENGTH	OPTICAL EFFICIENCY: 4×10^{-3} (1064 nm) 2×10^{-2} (532 nm) 8×10^{-2} (355 nm) SPECTRAL BANDWIDTH: 0.1 nm

In Table 1, preliminary estimates of ER-2 lidar system parameters necessary to support DIAL missions are shown. These estimates were generated by assessing the scientific mission requirements for the DIAL technique, and exercising an atmospheric simulation model for the DIAL measurement to infer minimum performance parameters for the instrument. Significant parameters to note are the modest energy per pulse and repetition rates, but the rather stringent requirement on the line width ($\leq$ 1.0 pm). The latter requirement translates into a stringent requirement for the wavelength identification and control system to be discussed later. In addition, it has been determined, through the atmospheric simulation model, that high precision and high accuracy measurements of water vapor in the atmosphere require that the amplified spontaneous emission emitted by the laser transmitter be less than 1 percent. This places a requirement to maintain a spectral purity in the emitted dye laser wavelengths to a spectral purity in excess of 99 percent. This in turn places requirements on the near and far field spatial pattern of the doubled Nd:YAG laser at 532 nm which may exceed commercial state-of-the-art capability.

Some of the salient performance characteristics for the ER-2 lidar instrument are summarized below.

- One Nd:YAG laser required to pump one-dye laser for differential absorption lidar (DIAL). Two sequential wavelengths required, spaced 400 μs apart.
- Amplified spontaneous emission (ASE) emitted by the dye laser shall not exceed 1 percent.

- Laser linewidth at 724 nm and 940 nm shall be < 1.0 pm.
- A totally autonomous instrument package within the envelope of the ER-2, with a benign vibrational environment, but a difficult thermal environment.

These functional requirements for the ER-2 lidar experiment can be translated into a series of performance requirements for the Nd:YAG pump laser. Salient performance requirements for the Nd:YAG pump laser are given in Table 2.

Table 2. Nd:YAG PUMP LASER PERFORMANCE REQUIREMENTS

• ENERGY PULSE:	300 - 350 mJ @ 532 nm (LONG TERM STABILITY) $\leq$10% (2σ) FROM SHOT TO SHOT
• PULSE PAIR SEPARATION:	400 μs $\pm$ 10 μs (2σ)
• SPECTRAL WIDTH:	$\leq$ 4 cm^{-1}
• LIFETIME:	> 1 x 10^6 PULSE PAIRS
• MAXIMUM INTENSITY:	< 200 Mw/cm^2
• INTENSITY PROFILE:	WITHIN 20% (2σ) OF A GAUSSIAN
• WAVELENGTH PURITY:	$\leq$ 1% OF 1064 nm PULSE ENERGY PRESENT AT 532 nm
• POLARIZATION:	LINEAR POLARIZATION $\geq$ 98:1

In order to obtain the requirements for high spectral purity for the double-pulsed dye laser transmitter, some rather difficult performance specifications are required for the spatial characteristics of the output spatial beams at 532 nm. Table 3 lists some of the salient beam characteristics of the Nd:YAG laser, as currently understood. It should be noted that at 0.74 m from the exit plane of the pump laser, a cylindrical lens is assumed to pump the dye laser; in the spatial distance from 0.74m to 4.0 m, the dye laser components which include dye laser cells and transfer optics are located.[+]

In addition to these laser performance requirements, difficult engineering requirements are imposed on angular and translational alignment stability and alignment adjustments on the laser output beam optics for pre-flight readiness, test and evaluation prior to autonomous operations on the ER-2 flight missions.

Studies are currently being conducted to further define the requirements and performance characteristics for the doubled Nd:YAG/dye laser transmitter for the LASE project, prior to prototype development and testing in the laboratory. As part of the technology development program, studies will also be conducted to understand the performance characteristics of a doubled ND:YAG laser, pumping a tunable solid state crystal. Results of these studies will be used to assess whether the requirements imposed upon the pump laser by the dye laser on the ER-2 aircraft are significantly relaxed by tunable solid state crystal technology. It is expected that these studies will significantly impact the technology selected for the ER-2 lidar program, and will provide valuable understanding of the tunable solid state laser technology to be advocated for future experiments from space platforms and Space Shuttle.

[+]The design of the dye laser is being conducted by the French National Institute of Astronomy and Geophysics (INAG).

Table 3. ND:YAG PUMP LASER SPATIAL BEAM CHARACTERISTICS

ND:YAG PUMP LASER SPATIAL BEAM CHARACTERISTICS
BEAM CROSS-SECTION AT 0.74 m

- SINGLE TRANSVERSE MODE (TEM_{00}) WITH DIFFRACTION LIMITED DIVERGENCE (DESIGN GOAL)
- CIRCULAR IN CROSS-SECTION, 0.74 TO 10 METERS FROM EXIT PLANE (DESIGN GOAL)
- BEAM CROSS-SECTION AT 0.74 m TO BE CIRCULAR WITH NOMINAL DIAMETER 8.5 - 9.5mm (DESIGN GOAL)

  OR

  ELLIPTICAL BEAM WITH ELLIPSE CONTAINED IN ANNULAR REGION DEFINED BY TWO CIRCLES OF 8.5 TO 9.5 mm DIAMETER.

BEAM CROSS-SECTION FROM 2.4 TO 4.0 M

- BEAM AT EACH CYLINDRICAL LENS SHALL BE CIRCULAR (DESIGN GOAL)

  OR

  BEAM AT EACH CYLINDRICAL LENS SHALL BE ELLIPTICAL. THE BEAM SHOULD SPREAD UNIFORMLY IN ALL ORTHOGONAL DIRECTIONS; BEAM MUST GROW UNIFORMLY IN CROSS-SECTION; BEAM NEED NOT GROW LINEARLY PROPORTIONAL TO DISTANCE.
- RECTANGULAR OR SQUARE CROSS-SECTIONS ARE UNACCEPTABLE
- DOUGHNUT SHAPED CROSS-SECTION ACCEPTABLE IF:
  - OUTSIDE DIAMETER OF RING IS SIMILIAR TO CIRCULAR BEAM REQUIREMENTS
  - OBSCURED CENTRAL SPOT SHALL BE $\leq$ 10% OF TOTAL AREA.
- BEAM SPREAD AT $<$ 10 METERS FROM EXIT PLANE SHOULD HAVE MAXIMUM BEAM SPREAD $<$ 0.6 mr, FULL ANGLE AT $1/e^2$ POINTS.

In parallel with these conceptual studies, alexandrite laser technology is also being assessed by research personnel at the Goddard Space Flight Center [GSFC/Dr. John Degnan], for two experiments scoped in Figure 3; namely the experiments required at 721 nm for measurements of lower tropospheric water vapor, and at 760 nm for measurements of pressure and temperature. Several presentations at this conference have addressed the status of alexandrite laser technology; interested readers should refer to these papers. Alexandrite lasers appear to satisfy most of the ER-2 lidar experiment requirements at 721 and 760 nm. Several technology advances still appear important for ER-2 experiment needs. First, development of a single mode laser beam with energy-per-pulse in the 100 mJ/pulse range appears desirable vis-a-vis the current multiple mode capability. In addition, extension of the current tuning capability to the 940 nm range through Raman shifting would provide in one laser technology the capability to conduct the major scientific experiments currently considered for the ER-2 lidar missions.

4. TUNABLE SOLID STATE LASER TECHNOLOGY

In developing long term plans for tunable solid state laser technology to meet needs for either a spaceborne or an ER-2 lidar mission, several critical technologies must be investigated, prior to commitment of a specific laser technology to flight systems. In these conference proceedings, several papers cover the status of four major activities in tunable solid state lasers including theoretical modeling of laser crystal materials, crystal growth techniques, characterization techniques of laser crystals, and the status of several operating tunable solid state lasers. For acceleration of this laser technology to meet potential missions for NASA needs, an interdisciplinary research program must be pursued where these four activities are closely coupled. Two other technology elements that must also be pursued aggressively are 1) wavelength shifting techniques to wavelengths currently unattainable directly with vibronic lasers; and 2) autonomous wavelength calibration and control techniques for ground and flight systems. Several concepts for laser wavemeters have been developed and tested in the laboratory.[12] For well-behaved, single-mode tunable lasers, these techniques appear adequate. However, for lasers with multiple modes and unstable mode amplitudes, further research is required, particularly as they relate to autonomous flight systems.

In Figure 4, a tunable solid state laser technology program currently being sponsored within the NASA Office of Aeronautics and Space Technology (OAST) is shown. Under the fundamental research element, studies are being initiated in theoretical modeling to understand fundamental processes and mechanisms that lead to the energy-per-pulse and average powers required for spaceborne lidar experiments. Other research activities include characterization of laser crystals and crystal growth techniques, with primary emphasis on those technologies that have potential for extension into the 5-10 joules per pulse range.

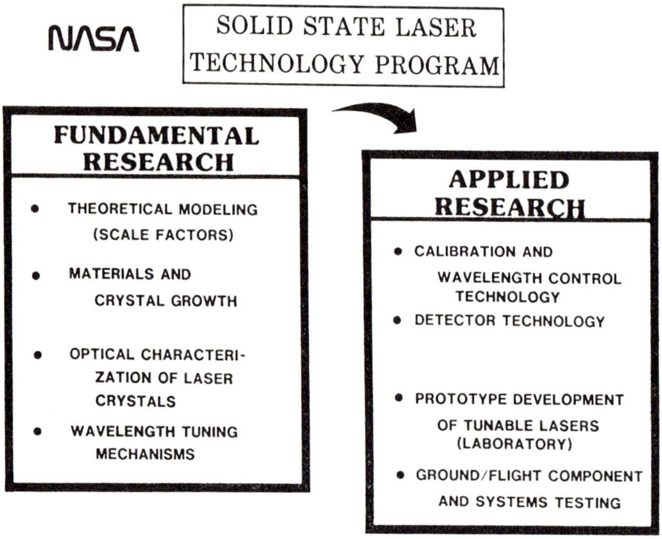

Figure 4

Under the applied research element, studies and laboratory measurements are being pursued in techniques for calibration and wavelength control, including relevant detector technology to cover the tuning range of solid state lasers. Ultimately, laboratory prototypes will be developed, with eventual testing in the laboratory and on the ER-2 lidar test bed. The ultimate objective of this program thrust is to provide an early assessment of tunable solid state laser technology for spaceborne lidar missions, currently being considered for Space Station observing platforms, free flying satellites, and the Space Shuttle.

5. CONCLUDING REMARKS

In considering the technological requirements for spaceborne missions with laser systems, there are several lidar instrument drivers for tunable solid state lasers (in addition to individual experiment requirements which dictate the laser wavelengths, energy-per-pulse, average power, and laser stability). These instrument drivers include a technology requirement for long lifetime (10^9-10^{10} shots), high reliability, high wall plug efficiency (>5%), non-cryogenic temperature operation, and low voltage operation. The focus at this conference has been primarily in the technology of vibronic lasers. However, there is a complementary technology that appears to satisfy the system performance requirements for spaceborne lidar missions and, when coupled to the technology of vibronic lasers, could satisfy the future system requirements for an "all solid state" tunable laser technology. The development of AlGaAs diode arrays as pump sources for Nd:YAG and Nd:GLASS rods or slab configurations has progressed at a rapid rate over the past several years.[13] At the 1060 or 532 nm wavelengths, this laser system with a high frequency stability and low chirp represents an alternative technology to CO_2 lasers for measuring two-dimensional winds in the earth's troposphere. At 532 nm, this laser system is also attractive as a pump source for some of the vibronic lasers discussed at this conference. In particular, Ti:Al_2O_3 pumped by a Nd:YAG or Nd:GLASS diode-pumped laser appears to be an attractive "all solid state" tunable laser system which incorporates many of the system performance characteristics necessary for spaceborne missions.

Tunable solid state laser systems configured with an "all solid state laser" pump as the basic optical pump source and used to pump a variety of solid state tunable crystals could provide multiple wavelength accessibility covering a wide range of the electromagnetic spectrum. In principal, this modular configuration has the potential to provide, either simultaneously or sequentially, a large number of wavelengths to execute lidar experiments from space. This capability is extremely important for the multidisciplinary requirements anticipated in future space mission.

For the future, current plans are to evaluate this technology through a combination of conceptual system studies and fundamental research activities. Alternative tunable solid state laser technologies will also be addressed, within this configuration of the "all solid state" pump as the primary building block, leading to a modular lidar instrument technology.

Acknowledgements

The authors would like to acknowledge members of project "LASE" who contributed to this paper. In particular appreciation is expressed to J. B. Talbot, project manager, W. M. Hall, instrument project engineer, Dr. E. V. Browell, project scientist, and G. W. Grew for providing the analysis of the computer simulation. In addition, R. R. Nelms and R. V. Hess of the Flight Electronics Division at Langley provided valuable information toward the current technology thrusts of the lidar research program. A special thanks goes to Malinda P. Boyce for assembling the final manuscript.

References

1. "Space Station: Policy, Planning and Utilization": AIAA Aerospace Assessment Series, Vol. 10, Proceedings of the AIAA/NASA Symposium on the Space Station, Arlington, Virginia, July 18-20, 1983.

2. Earth Observing System is a NASA Sponsored Study Collaborative Between the Goddard Space Flight Center and the Jet Propulsion Laboratory. The Program Scientist is D. Butler (NASA Hdqtr's); GSFC Manager (C. Mackenzie).

3. "Airborne and Ground-Based Lidar Measurements of the El Chicon Stratospheric Aerosol from 90 ^{O}N to 56 ^{O}5": McCormick, M. P., et al; to be published in Geofisica International, June 1984.

4. "Lidar Remote Sensing of Tropospheric and Trace Gases - Programs of NASA Langley Research Center": Browell, E. V.; Proceedings of the Fourth Joint Conference on Sensing of Environmental Pollutants, American Chemical Society, 395-402.

5. "Shuttle Atmospheric Lidar Research Program", Final Report of Atmospheric Lidar Working Group, NASA SP-433, 1979.

6. "Atmospheric Lidar Multi-User Instrument Definition Study", (ed., R. V. Greco), NASA Contractor Report 3303, August 1980.

7. "The Global Troposphere: Biogeochemical Cycles, Chemistry, and Remote Sensing": Levine, J. S. and Allario, F.; Environmental Monitoring and Assessment 1 (1982), 263-306.

8. Upper Atmosphere Research Satellite Program, "Final Report of the Science Working Group": JPL Publication 78-54, July 15, 1978.

9. "A Second-Generation Laser Radar": Kent, G. S., Sandland, P., Wright, R. W. H.; Jl. Appl. Met., Vol 10, 443-452, 1971.

10. "An Airborne Lidar System for Ozone and Aerosol Profiling in the Troposphere and Lower Stratosphere", in J. London (ed.), Proceedings of the Quadrennial International Ozone Symposium, 99-107.

11. "Lidar Measurements of Tropospheric Gases": Browell, E. V.; Opt. Eng.21, 128-132.

12. Fizeau Wavemeter: Snyder, J. J.; Spie Vol. 288, Los Alamos Conference on Optics, P. 258 - 262, (1981).

13. "Phased Array Diode Lasers": Streifer, W., et al; Laser Focus, Vol. 20, No. 6, Pages 100-109, June 1984.

Laser Transmitters for Meteorological Lidar: NTAL Program

F.P. Roullard, III

Northrop Electronics Division[1], Hawthorne, CA 90250, USA

1 Abstract

A DIAL lidar technique using narrow-band tunable alexandrite laser (NTAL) transmitters is presented as a means of acquiring meteorologically important atmospheric properties. Performance specifications are given for NTAL-1 (1981) and NTAL-2 (1983). The optical train of the NTAL transmitters is described. In band spectral distribution measurement results are shown. Out of band spectral content is estimated and an experimental upper bound determined.

2 Introduction

The narrow-band tunable alexandrite laser (NTAL) program has so far consisted of two NTAL units which Allied has delivered to NASA Goddard Space Flight Center. The first unit was delivered in April 1981 [1,2], the second in June 1983 [3]. Both lasers are now part of a DIAL lidar experiment for remote sensing of meteorologically important atmospheric properties.

3 Meteorological Lidar

A DIAL lidar experiment using the NTAL-1 unit has already been completed, obtaining atmospheric pressure and temperature profile measurements at O_2 absorption wavelengths, by C. L. Korb et al at the NASA Goddard Space Flight Center [4]. Simulations predict a 0.3% accuracy for pressure profile measurements using an alexandrite lidar system modified for use on the space shuttle [5]. Since alexandrite can also be tuned to the H_2O absorption line near 724 nm, lidar humidity measurements can be made using this same laser system [6]. Figure 1 shows how the alexandrite tuning range overlaps the spectral lines which are important for meteorological lidar, using the differential absorption lidar (DIAL) technique. The tunability of the alexandrite laser provides a means to discriminate between the molecular absorption of H_2O or O_2 versus the background scattering of atmospheric aerosols.

The absorption bands of Figure 1 are each a manifold of several dozen individual lines, a few of which have optimum characteristics for the LIDAR measurement. Figure 2 illustrates the spectral technique useful for pressure measurements. The laser wavelength is tuned to the absorption minimum point between two pressure-broadened oxygen lines at 769 nm. Pressure differences cause absorption differences, as indicated by the plots for surface and 4 km altitude. The slow variation versus wavelength make

[1]Work was done at Allied Technologies, Military Laser Products, 31717 La Tienda Drive, Westlake Village, CA 91362.

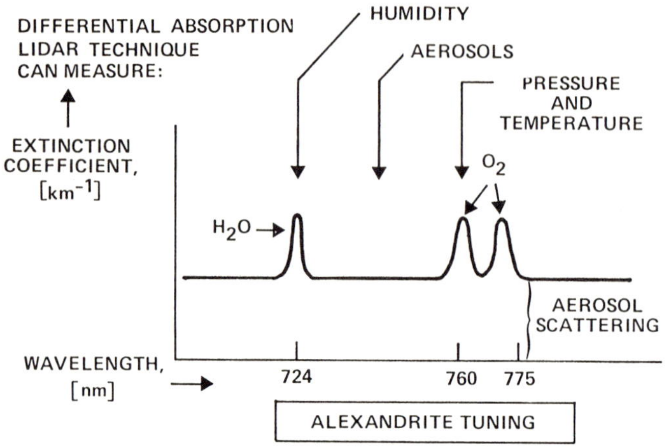

Fig. 1 Alexandrite tuning range and spectral lines for meteorological lidar

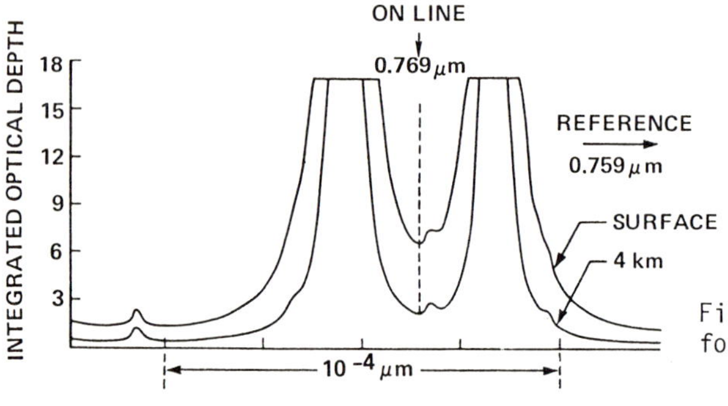

Fig. 2 Spectral technique for pressure measurements

the measurement less sensitive to drift of the laser wavelength. Aerosol scattering is compensated by conducting a simultaneous attenuation measurement at 759 nm. For temperature measurement, a line that is sensitive to changes in the thermal Boltzman distribution is employed, such as the 768 nm line. The laser wavelength is adjusted to the absorption maximum in this case. Humidity is measured in a similar fashion by the strength of the line at 724 nm.

The DIAL lidar system is schematically illustrated in Figure 3. The system consists of the NTAL transmitters, as represented by the NTAL-2 unit, a collection of detector modules, as represented by the spectrometer and telescope, and the computerized data-acquisition subsystem. Note that the remote control panel of the NTAL-2 is installed in the same rack as the data-acquisition subsystem. Complete control of the laser wavelength is provided by this panel; no internal access to the laser is required for any of the tuning elements during normal operation.

4 The NTAL Laser Transmitters

The performance specifications for the two NTAL units are presented in Table 1. The most significant differences between the two lasers are the specifications relating to fine tuning. The NTAL-2 unit provides improved

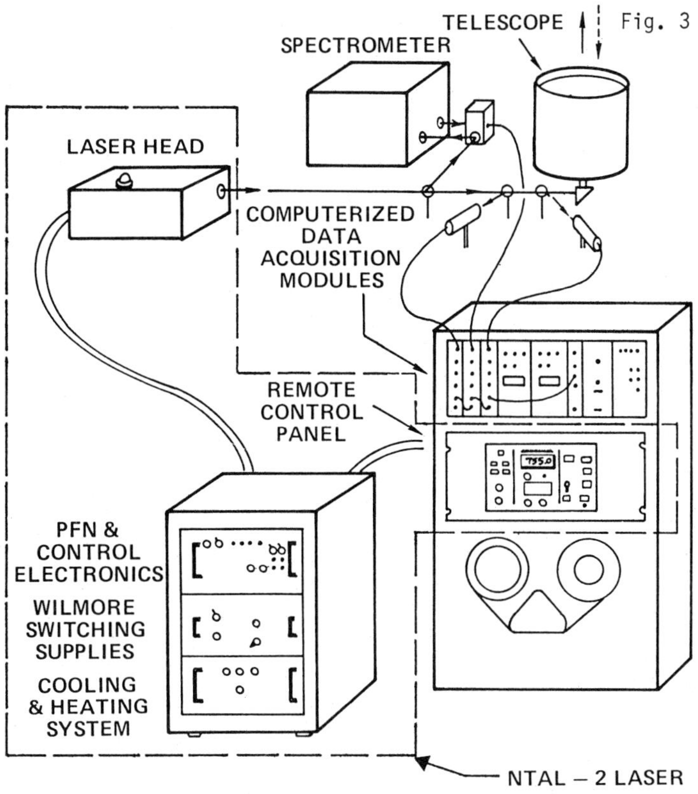

Fig. 3 DIAL lidar system

performance as a result of the remote electronic gang tuning of the tuning elements.

Table 1 Performance Specifications

	NTAL-1 (1981)	NRAL-2 (1983)
Repetition Rate [Hz]	10	10
Pulse Energy [mJ]	100	150
Pulse Duration [ns]	200	200
Tuning Range [nm]	750-770	740-775
Spectral Bandwidth [cm^{-1}]	0.02	0.02
Spectral Stability [cm^{-1}/30 Min.]	±0.01	±0.01
Fine Tuning Range [cm^{-1}]	0.5	2.5
Fine Tuning Method	Temperature Tuning of Thick Etalon While Manually Adjusting the Birefringent Filter and Thin Etalon to Avoid Mode Hopping and Maintain Power	Remote Electronic Gang Tuning

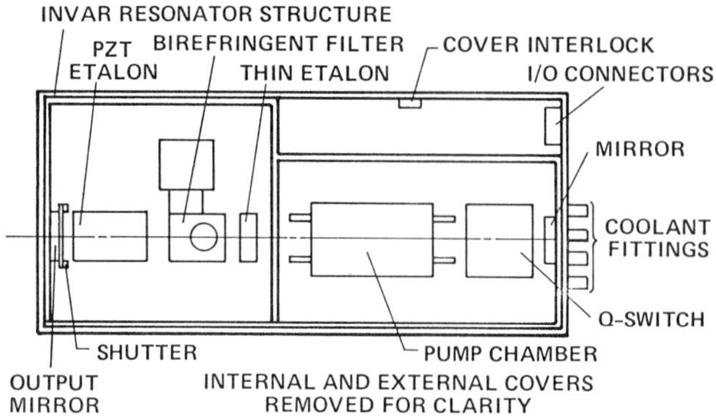

Fig. 4 Optical Train

The NTAL-2 optical train is shown in Figure 4. A ruggedized invar resonator structure supports the two end mirrors, the tuning elements, and the Q-switch. The pump chamber is mounted upon a separate base plate to reduce vibrational effects due to the cooling loop.

The passbands of the three tuning elements are illustrated in Figure 5. The birefringent filter is adjusted to select a single passband of the tilt tuned etalon. The tilt tuned etalon is in turn adjusted to select one passband of the piezoelectrically tuned etalon. All three tuning elements are electro-mechanically controlled, so that all tuning adjustments are performed by potentiometers on the remote control panel. The tuning circuitry provides synchronous gang tuning of all three elements over the specified 2.5 cm^{-1} fine tuning range. This range was chosen as being needed to assure proper identification of spectral location. Note that in Figure 2, a one Angstrom (10^{-4} micron) interval is illustrated; at these wavelengths that interval is equivalent to a 1.7 cm^{-1} scan range.

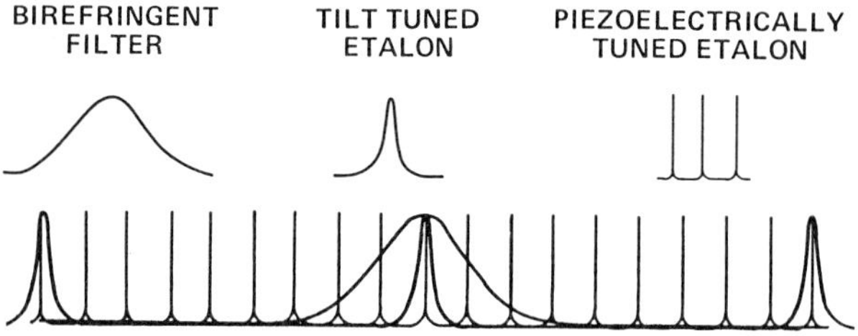

Fig. 5 Passbands of the three tuning elements

5 Spectral Characterization of Laser Emission

A pulse laser spectrum analyzer for measuring the in-band spectral content is shown in Figure 6 [7]. The laser beam enters a collimator and then passes through an interferometer. The resulting pattern is focused upon the multi-element photodiode array by means of a cylindrical lens. The interferometer is adjusted to provide a one wavelength wedge over the width of the beam, producing images of three passbands upon the array detector.

The photodiode array output for a single shot typical of NTAL-1 performance is shown in Figure 7. The pattern consists of three modes whose

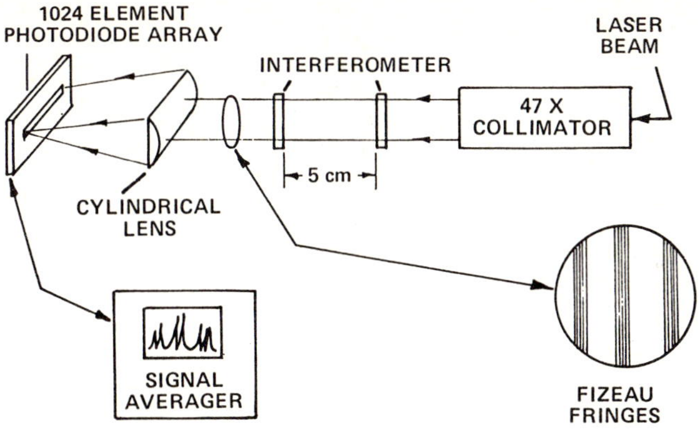

Fig. 6 Pulse laser spectrum analyzer

relative spacing is asymmetric, the spacings being 0.011 cm^{-1} and 0.015 cm^{-1}. This discrepancy is evidently due to the presence of higher order transverse modes, the existance of which is confirmed by comparing the size of the TEM_{00} beam observed at threshold to the two or three times larger beam size seen at the specified output energy level.

The out of band spectral content has a potentially critical influence on the success of a meteorological lidar measurement scheme. As stated by F. Allario, the out of band emission in a dye laser-based lidar system is about 1% and represents a severe problem [8]. The out of band emission can be simply estimated in the case of alexandrite, since the gain of the medium is low and losses at undesired wavelengths are high due to the use of three tuning elements. Assume that the double pass gain is equal to the single pass loss through the tuning elements and output coupler, and that the laser is pumped at the maximum input level of 200 J and has a storage efficiency of 4%. The laser is typically operated near threshold at 150 mJ. The relatively long pulse build up time near threshold provides several hundred passes through the tuning elements, thus giving the desired narrow line width while incurring the expense of an efficiency less than 0.1%, which is a non-critical issue for this application. Under these conditions, an estimate of the broadband fluorescence energy contained within the 3 milliradian output beam is then 4 micro-Joules, giving a 30 parts per million ratio of out of band energy to in band energy. C. L. Korb and G. Schwemmer have recently determined experimentally an upper bound of 100 parts per million [9]. This measurement is in good agreement with the above estimate.

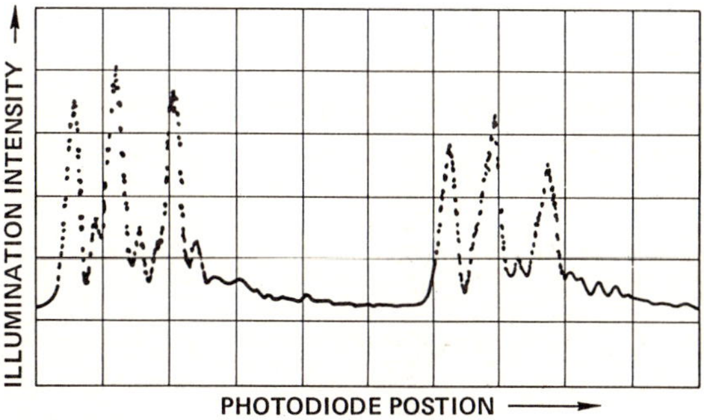

Fig. 7 Photodiode array output for a single shot

6 Conclusion

The NTAL program has so far demonstrated that alexandrite lasers are suitable for use in meteorological lidar. This utility is based upon performance beyond the mere coincidence of laser wavelength range and molecular spectral absorption bands. The ease of tunability and capability for narrow bandwidth operation have been demonstrated. The low out of band spectral emission is a feature that promotes alexandrite above competing laser types. These features justify the interest of Allied and NASA in considering alexandrite for future programs.

7 Acknowledgements

The NTAL-1 project was directed by R. Sam; H. Marcus performed the mechanical design. Most of the major subcomponents of the NTAL-2 were designed and built by the staff at Laser Photonics, including R. McKinney and D. Zander. During system integration of NTAL-2, A. Schloss provided electronic design assistance and R. Smith provided mechanical design assistance. Much appreciation is due M. Dombrowsky, L. Korb, and G. Schwemmer at NASA Goddard Space Flight Center for their encouragement of the NTAL program and for contributing their test results of the delivered laser systems.

References

1. R. C. Sam, F. P. Roullard, III, "Narrow-band tunable alexandrite laser for meteorological lidar," E.O.S.D., Vol. 14, pp. 37-40, July 1982.

2. Laser was delivered under Contract No. NAS5-26300.

3. Laser was delivered under Contract No. NAS5-2491-003.

4. C. L. Korb, G. K. Schwemmer, M. Dombrowski, "Atmospheric pressure and temperature profile measurements using near IR lidar systems," Workshop on Optical and Laser Remote Sensing, Feb. 1982, Monterey, CA, sponsored by U. S. Army Research Office.

5. C. L. Korb, C. Y. Weng, "Differential absorption lidar technique for measurement of the atmospheric pressure profile," Appl. Opt., Vol. 22, pp. 3759-3770, Dec. 1983.

6. T. D. Wilkerson, G. Schwemmer, "Meteorological lidar developments," Proc. Los Alamos Conference on Optics '81, SPIE Vol. 288, pp. 383-386, Apr. 1984.

7. G. K. Schwemmer et al, Eleventh International Laser Radar Conference, Univ. of Wisconsin-Madison, June 21-25, 1982.

8. F. Allario, in this volume.

9. C. L. Korb, G. Schwemmer, private communications.

Part II

Vibronic Laser Materials Spectroscopy

Recent Results on the Spectroscopy of Transition Metal Ions for Tunable Solid State Lasers

Richard C. Powell

Department of Physics, Oklahoma State University, Stillwater, OK 74078, USA

I. Introduction

The recent expansion of electro-optic based technology has produced a demand for laser systems having a variety of different operational parameters. In many cases, solid state lasers are the most desirable choice because of their ruggedness and minimum maintenance requirements. Because of this, there is significant interest in research and development on optical materials for use as solid state lasers. Materials which exhibit broad band vibronic emission are of special importance in developing lasers whose wavelength can be tuned continuously over a broad spectral range. Research in this area can be divided into categories: searching for new vibronic laser materials and improving existing vibronic laser materials. This paper summarizes the results obtained so far on several projects currently underway in our laboratory which fall into one or the other of these two categories.

Modern optical spectroscopy techniques provide powerful methods for studying the fundamental properties of a material which are important in characterizing its potential for use in solid state laser systems. The development of different types of laser spectroscopies allows us to probe the properties of a sample with much higher spatial, spectral, and temporal resolution than ever before available as well as allowing us to monitor the response of a material to high power laser radiation and coherent driving fields. The information gained from these experiments can be correlated with crystal growth parameters to help develop improved material quality and in addition can be correlated with laser operational parameters to identify physical reasons for device limitations. The use of several different experimental methods to completely characterize all aspects of the spectral dynamics of a material is important in determining its optimum properties for device applications. A variety of techniques are currently being used in our laboratory including time-resolved site-selection spectroscopy, four-wave mixing spectroscopy, picosecond spectroscopy, nonlinear spectroscopy, and Raman microprobe spectroscopy. Several of these were used in the work described below.

In the following sections, three different projects are discussed. The first involves a survey of the spectroscopic porperties of 4d and 5d transition metal ions in different types of host crystals. The second concerns the use of Raman microprobe techniques to identify scattering centers in Al_2O_3:Ti^{3+} crystals. The final project involves measuring the four-wave mixing properties of alexandrite crystals.

II. Spectroscopic Survey of 4d and 5d Transition Metal Ions

The currently operating vibronic lasers are all based on 3d transition metal ions usch as Cr^{3+}, Ti^{3+}, Co^{2+}, and Ni^{2+} in various types of host crystals.

DIVALENT & TRIVALENT TRANSITION METAL IONS

$3d^{1(9)}$	$3d^{2(8)}$	$3d^{3(7)}$	$3d^{4(6)}$	$3d^{5}$
Sc^{2+}	V^{3+}	V^{2+}	Mn^{3+}	Mn^{2+}
Ti^{3+}	(Ni^{2+})	Cr^{3+}	(Fe^{2+})	Fe^{3+}
(Cu^{2+})	Ti^{2+}	(Co^{2+})	(Co^{3+})	
		(Ni^{3+})	Cr^{2+}	

$4d^{1(9)}$	$4d^{2(8)}$	$4d^{3(7)}$	$4d^{4(6)}$	$4d^{5}$
Y^{2+}	(Pd^{2+})	(Rh^{2+})	(Ru^{2+})	Ru^{3+}
Zr^{3+}		(Pd^{3+})	(Rh^{3+})	
(Cd^{3+})				

$5d^{1(9)}$	$5d^{2(8)}$	$5d^{3(7)}$	$5d^{4(6)}$	$5d^{5}$
La^{2+}	(Pt^{2+})	(Ir^{2+})	Re^{3+}	Re^{2+}
		(Pt^{3+})	(Os^{2+})	Os^{3+}
			(Ir^{3+})	

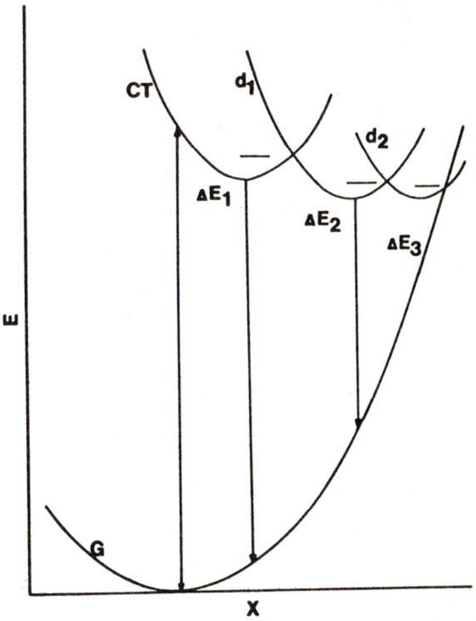

Fig. 1 d electron configurations for divalent and trivalent transition metal ions

Fig. 2 Simplified configuration coordinate model for interpreting the observed spectral dynamics of $4d^n$ and $5d^n$ transition metal ions in crystals

Another class of ions which have similar unfilled d-electron configurations are the 4d and 5d transition metals. Figure 1 lists the ions of this type in terms of their electron configurations. Under certain conditions, chemical complexes incorporating these ions are known to exhibit efficient broad band fluorescence emission as desired for tunable laser applications. However, little work has been done on studying the spectroscopic properties of these ions as dopants in host crystals. We are presently carrying out an investigation to evaluate the potential of these ions in solid state laser materials. So far, divalent ions of Ru, Rh, Ir, and Pt have been studied and found to have similar spectral properties [1]. In this paper, the spectroscopic properties of Rh^{2+} in three different host crystals are compared as an example of the results obtained in this project.

The 4d and 5d transition metal ions of interest to this work have greater than half filled shells. The d^5 through d^8 electron configurations all have low lying excited states strongly coupled to the crystal field of the host. Since the 4d and 5d ions have strong spin-orbit coupling, a simple crystal field analysis of the d^n states does not provide as accurate a description of the energy levels of these ions in crystals as it does for $3d^n$ ions [2]. However, it is generally true that there are several excited levels whose energy with respect to the ground state increases significantly as the crystal field increases and whose transitions to the ground state are either dominantly spin-allowed or spin-forbidden. These properties are favorable for broad band vibronic emission. Figure 2 shows a schematic configuration coordinate diagram for this type of ion. Strong charge transfer (CT) transitions dominate the absorption. After excited state relaxation, either the energy can be emitted from this charge transfer state or intersystem crossing to the d levels can occur. Since there are several different d levels, the energy is distributed among these after which both radiative and radiationless emission occurs. The fluorescence lifetimes of the CT, spin allowed d, and spin forbidden d transitions are of the order of nanoseconds, microseconds, and milliseconds, respectively. This variety of types of levels and transitions results in complex spectral dynamics for these ions.

For tunable laser materials, it is desirable to have strong absorption bands and broad emission bands with high peak cross-sections. The latter quantity varies directly with quantum efficiency and inversely with the band width and lifetime. Thus the ideal operating scheme for these ions would be to pump into the charge transfer band and have efficient cross over of the excited state energy to the d levels having spin-allowed transitions to the ground state with minimum radiationless loss. The CT absorption for each of these ions in all of the host crystals investigated so far was found to occur in the near uv spectral region. The CT absorption bands are strong, broad, and structured with maxima occurring near 320 nm. This makes them ideal for pumping with a N_2 laser at 337.1 nm. The most important question is, which host crystals maximize cross over to the d levels for fluorescence emission?

The first set of host crystals investigated were the alkali halides because of our ability to grow them easily. These were grown by the Czochralski method in a dry nitrogen atmosphere. Good optical quality single crystals were obtained which were cleaved from the boule perpendicular to the growth axis to ensure uniform doping. Time-resolved fluorescence spectra were obtained as a function of temperature to observe the spectral dynamics. The experimental apparatus employed N_2 laser pumping and boxcar integrator detection. The details of the system are described elsewhere [1].

Figure 3 shows the fluorescence emission from $KBr:Rh^{2+}$ at room temperature. The strong, sharp band peaking near 380 nm has a lifetime of the order of 20 ns and is thus attributed to CT emission. The broad, structured band extending throughout the visible spectral region has a lifetime of the order of 25 μs and is attributed to emission through spin-allowed d-d transitions. As temperature is lowered the fluorescence lifetime of the d emission increases to about 80 μs at 10 K while the relative intensity ratio of the d emission to the CT emission increases down to about 250 K and then decreases as tem-

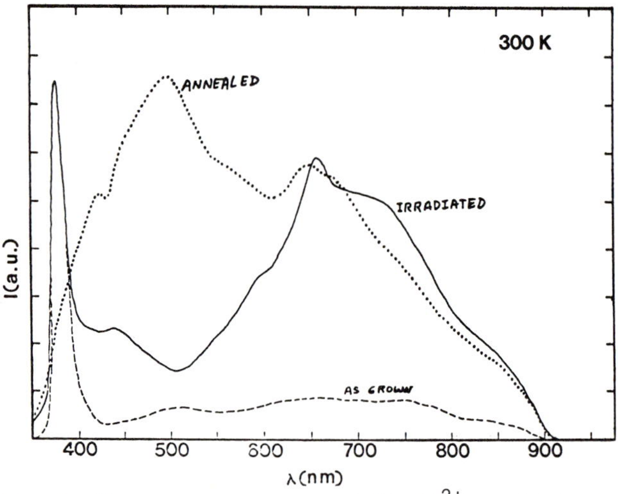

Fig. 3 Fluorescence spectra of $KBr:Rh^{2+}$ at 300 K after pulsed N_2 laser excitation. ----- untreated sample; ——— irradiated sample; · · · · · annealed sample

perature is lowered further to 10 K. Analysis of these data allow for the determination of the activation energy for intersystem crossing of the configuration coordinate diagram shown in Fig. 2.

After demonstrating that emission through spin-allowed d-d transitions can occur in this material, the next problem is enhancing the fraction of emission occurring in this mode compared to the CT emission. This can be done by shifting the position of the CT configuration coordinate compared to the configuration coordinates of the d levels in order to lower the potential barrier for crossover. Since the CT levels are strongly affected by changes in their ligand environments, the introduction of lattice defects such as vacancies as nearest neighbors to the Rh^{2+} ions should change the CT levels significantly. Two processes were attempted to accomplish this. The first was thermal annealing. The sample was heated to 600 C and quenched on a copper block. The second process was irradiating the sample with a dose of 10^5 rads of electrons from a 2 MeV van de Graaff accelerator. The spectra obtained after each of these treatments are shown in Fig. 3. Both treatments were successful in producing a relative enhancement of the d emission. The radiation treatment results in a greater increase in the low energy part of the d band and the thermal treatment results in a greater increase in the high energy part of this band.

Table 1. $Rh^{2+}(4d^7)$ Spectroscopic Parameters

PARAMETER	HOST CRYSTAL		
	Na β"-Alumina	KBr	$LiNbO_3$
r[Å]	2.87	1.33	0.60
λ_p [nm]	700	650	568
$\Delta\lambda$ [nm]	130	270	72
τ_f [μs]	5200	25	0.577
σ_p [$10^{-21} cm^2$]	0.047	2.10	627

The spectral parameters obtained in this work are summarized in Table 1. r represents the size of the site occupied by the Rh^{2+} ion, which has an ionic radius of 0.8 Å. λ_p is the peak wavelength of the d emission band and $\Delta\lambda$ is the full width of the band at half maximum. τ_f is the room temperature fluorescence lifetime of the d emission band and σ_p is its peak emission cross-section. It can be seen from the table that KBr crystals provide a host for Rh^{2+} which gives a good possible tuning range and a cross-section only slightly smaller than 3d transition metal ions. The major problem is obtaining samples with increased dopant concentrations so that reasonable gain can be achieved in laser operation. Attempts to grow single crystals of alkali halides containing high concentrations of 4d and 5d transition metal ions have been unsuccessful.

In order to work with higher doping concentration, different types of host crystals were investigated. One interesting possibility is using a crystal having a layered structure and diffusing the doping ions into the regions between the layers instead of doping during the growth process. For this purpose, samples of $Na_{1.67}Mg_{0.67}Al_{10.33}O_{17}$ (β") were obtained from Oak Ridge National Laboratory. These Na β"-alumina crystals have layers with the spinal crystal structure bridged by Na^+ ions [3]. The spacing between the layers is about 4.76 Å. By heating this crystal for several hours in a moulton solution of NaCl-$RhCl_3$ with varying concentrations, it was possible to diffuse Rh^{2+} ions into the open planes and replace the Na^+ ions in concentrations of the order of 1 wt.%.

The Rh^{2+} absorption in β"-alumina crystals was again due to a CT band in the near uv region [4]. The fluorescence obtained at room temperature at two times after the N_2 laser pumping pulse are shown in Fig. 4. The broad fluorescence emission band has a lifetime of about 5.2 ms implying that the dominant emission is from a spin-forbidden d-d transition. This results in an extremely small peak emission cross-section as listed in Table 1.

$LiNbO_3$ crystals are another type of host for Rh^{2+} ions. As indicated by the results given in Table 1, this material produces a broad emission band peaking near 568 nm and having a fluorescence lifetime of 0.577 μs. This results in a very favorable peak emission cross-section. Since laser induced "optical damage" which produces light scattering and absorption is known to be a severe problem in $LiNbO_3$ crystals, they are not suitable for laser host materials. However, the spectral trends seen from the data listed in Table 1 may be helpful in identifying an appropriate host crystal for Rh^{2+} ions. As the parameter r decreases the crystal field strength at the Rh^{2+} ion site increases. The data show that the positions of the d emission band peaks shift to higher energy with increasing crystal field strength as predicted by the Sugano-Tanabe diagrams and the fluorescence lifetimes become shorter, indicating a shift from predominantly spin-forbidden to predominantly spin-allowed emission. Apparently an optimum host for producing the desired peak emission cross-section and accepting a high concentration of dopant ions will be one in which the Rh^{2+} ions replace ions having similar ionic radii. Various possibilities are currently being considered.

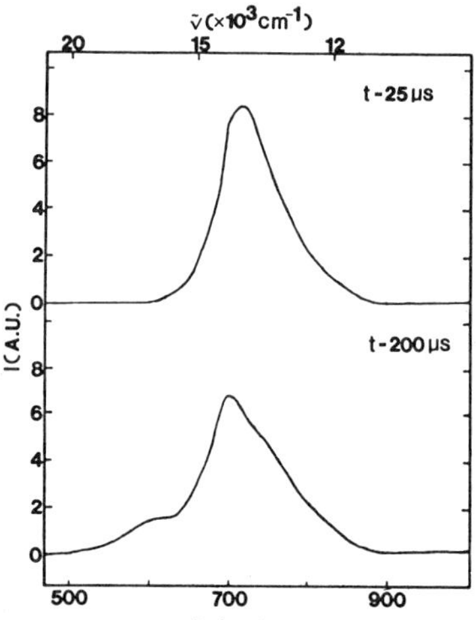

Fig. 4 Room temperature fluorescence spectra of Rh^{2+} in Na β"-alumina crystals at two times after pulsed N_2 laser excitation

III. Laser Raman Microprobe Analysis of Al_2O_3:Ti^{3+} Crystals

Tunable vibronic laser action has been obtained from Ti^{3+} ions in Al_2O_3 crystals and this material appears to be quite promising as a new tunable solid state laser [5]. One problem which has inhibited the development of Al_2O_3:Ti^{3+} lasers is poor crystal quality. Although new crystal growth methods for sapphire based on heat exchange techniques have greatly improved undoped host crystal quality, doped crystals suffer from the presence of small inclusions or bubbles. These scatter light and thus act as a loss mechanism when the material is used in laser applications. Thus it is important to identify the nature of these scattering centers so crystal growth parameters can be changed in such a way as to minimize their concentration.

Laser Raman scattering has been developed as a useful spectroscopic tool for identifying different chemical species. Used in conjunction with a microscope, laser Raman microprobe spectroscopy can be used to study micron size regions of a sample [6]. We recently applied this technique to the study of inclusions in $Al_2O_3:Ti^{3+}$ crystals using a Spectra Physics argon ion laser with an Instruments SA microscope, monochromator, and computer system [7].

The samples used in this work were grown by Dr. J. L. Caslavsky using the VSOM method. A significant concentration of scattering centers could be seen visually. Laser Raman microprobe spectra were obtained both in the normal region of the sample and by focusing the laser on one of the scattering centers.

In the perfect region of the crystal the Raman peaks coincided closely with those observed in undoped sapphire crystals and a strong background fluorescence from Ti^{3+} was detected. When the laser was focused onto a scattering center the results were significantly different. No background Ti^{3+} fluorescence was observed and the Raman peaks shifted to positions similar to those obtained on a sample of TiO_2. The latter spectra are shown in Fig. 5.

The results of this work indicate that inclusions of TiO_2 precipitates are responsible for the scattering centers in $Al_2O_3:Ti^{3+}$ crystals. Adjusting crystal growth parameters to minimize the formation of these precipitates has resulted in crystals with a greatly reduced concentration of scattering centers.

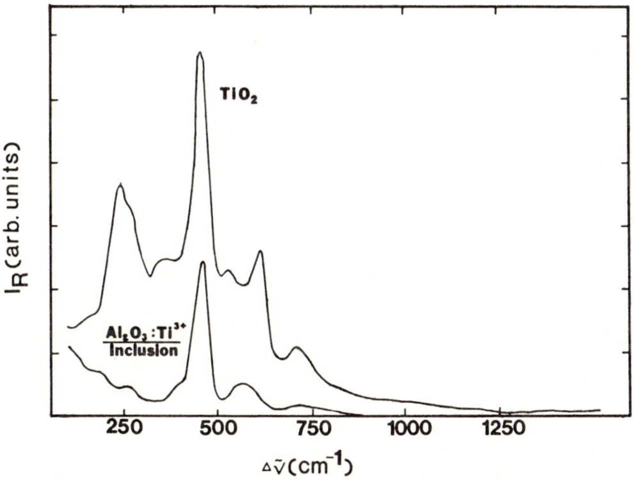

Fig. 5 Laser Raman microprobe spectrum of an inclusion in $Al_2O_3:Ti^{3+}$ compared to the spectrum of TiO_2 obtained under similar experimental conditions

IV. Four-Wave Mixing in Alexandrite Crystals

$BeAl_2O_4:Cr^{3+}$ (alexandrite) has been the most successful tunable solid state laser material to date [8]. To fully understand its operational limitation, it is important to understand all of the physical properties of the material. So far the nonlinear optical properties of alexandrite crystals have not been well characterized and we are currently using degenerate four-wave mixing techniques to do this [9].

The geometry of the experiment is shown in Fig. 6. Two beams from either an argon laser or an argon-pumped dye laser with R6G dye cross at an angle θ inside the sample in the ab plane with horizontal polarization. The interference of these write beams P_1 and P_2 produce a holographic grating in the sample. The probe beam P_r is aligned conjugate to one of the write beams and Bragg diffracts off the grating to give the signal beam S. By chopping the write beams off and monitoring the decay of the signal beam the dynamics of the grating decay can be studied.

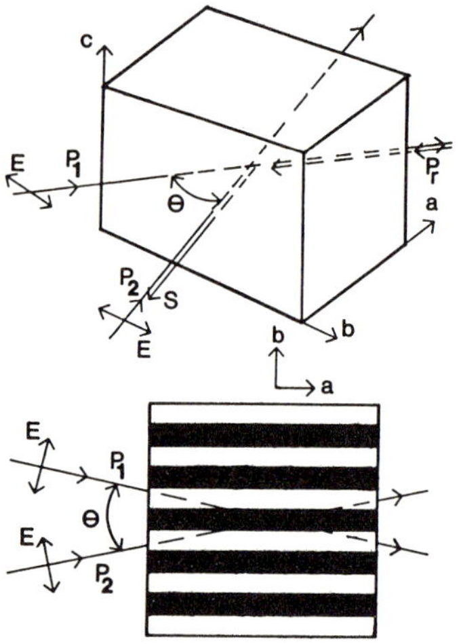

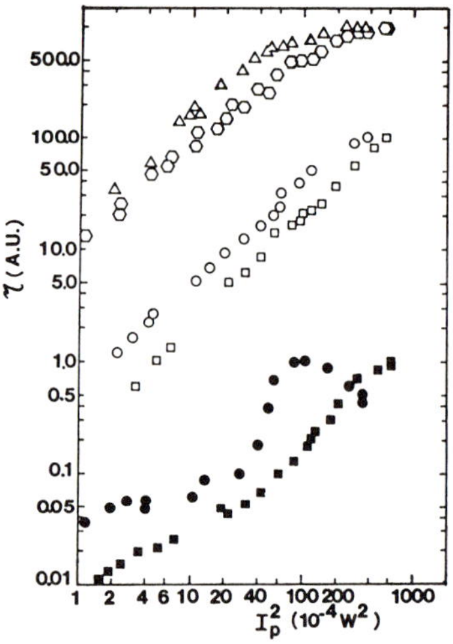

Fig. 6 Geometry of four-wave mixing experiment in alexandrite

Fig. 7. Power dependence of four-wave mixing efficiency in alexandrite Bulk: λ = 488 nm, Θ = 6° (○); λ=514.5 nm, Θ = 6° (△); λ = 488 nm, Θ = 26.5° (□); λ = 514.5 nm, Θ = 26.5° (○). Surface: λ = 488 nm, Θ = 26.5° (■); λ = 514.5 nm, τ = 26.5° (●).

We established and probed transient holographic gratings in an alexandrite crystal at three laser wavelengths, 488.0, 514.5, and 579.1 nm. The maximum phase conjugate scattering efficiency was approximately 10^{-3} for each of these wavelengths for the orientation shown in Fig. 6. It was possible to form gratings only in the b direction. By measuring the grating decay rate as a function of the crossing angle of the write beams, it was found that the grating decay was consistent with the decay of the Cr^{3+} ions from the ^2E level. This shows that the four-wave mixing is due to an excited state population grating and that no long range energy transfer among Cr^{3+} ions occurs. In addition, the results indicate that the 579.1 nm excitation selectively creates a grating with ions in the mirror sites, while the other two wavelengths selectively create gratings with ions in the inversion sites. Figure 7 shows the variation of four-wave mixing scattering efficiency with write beam laser power. In general a quadratic dependence is observed as expected. Deviations from this occur for gratings created near the crystal surface. Also at high powers a tendency toward saturation can be observed.

The direct importance of nonlinear optical effects such as those described here to laser performance is not yet clear.

V. Summary

The preceeding sections describe the use of time-resolved spectroscopy techniques to elucidate the spectral dynamics of 4d and 5d transition metal ions in crystals, laser Raman microprobe spectroscopy to identify crystal defects, and four-wave mixing spectroscopy to characterize nonlinear optical properties of crystals. These very different projects demonstrate how optical spectroscopy techniques can be used in vibronic laser research both in the search for new materials and in improving existing materials.

Acknowledgments

This work was sponsored by contracts from the Army Research Office and the Office of Naval Research.

References

[1] R.C. Powell, R.H. Schweitzer, J.J. Martin, G.E. Venikouas, and C.A. Hunt: J. Chem. Phys., to be published
[2] S. Basu and A.S. Chakravarty: Phys. Rev. B 26, 4327 (1982)
[3] J.B. Bates, J.C. Wang, and N.J. Dudney: Physics Today 35, 56 (1982)
[4] L. Xi, G.E. Venikouas, R.C. Powell, and J.B. Bates, to be published
[5] P.F. Moulton: Laser Focus 19, 83 (1983)
[6] G.J. Rosasco: Adv. in Infrared and Raman Spect. 7, 223 (1980)
[7] R.C. Powell, J.H. Bowen, A.W. Hounslow, and J.L. Caslavsky, to be published
[8] J.C. Walling: Laser Focus 18, 45 (1982)
[9] A.M. Ghazzawi, J.K. Tyminski, R.C. Powell, and J.C. Walling, to be published

Measurements of $Ti^{3+}:Al_2O_3$ as a Lasing Material

Georg F. Albrecht, J.M. Eggleston, and J.J. Ewing

Mathematical Sciences Northwest, Inc., 2755 Northup Way
Bellevue, WA 98004, USA

Broadly tunable solid state lasers[1-11] have long been of interest for a variety of applications in and out of the laboratory. Requirements with respect to pulse energy, beam quality and possibly high average power operation however put additional demands on the active medium. In addition to tunability a practical active medium must be growable to sufficient crystal size with an optical quality that allows for diffraction limited applications, permits room temperature operation and has a suitable cross-section. An insufficient cross-section ($\lesssim 10^{-20}$ cm^2) corresponds to a saturation fluence approaching or exceeding optical damage thresholds, and consequently will not allow efficient extraction of the active medium at the short pulse durations necessary to efficiently employ nonlinear frequency conversion schemes which greatly extend the range of tunability.

Together with other investigators[8,12,13] we believe that $Ti^{3+}:Al_2O_3$ is an active medium with great promise and therefore have investigated its lasing properties in our laboratory. Since Ti^{3+} has only one d electron, its Tanabe Sugano diagram in an octahedral environment is very simple. (See Fig. 1.) No upper state absorption into other d electron states is possible and no intermediate metastable level lengthens the upper state lifetime thus reducing the stimulated emission cross-section as is the case in medium to high field Cr^{3+} doped crystals[14,15]. The short upper state lifetime [$\tau(1/e) = 3.2$ μsec] makes pumping with conventional flashlamps inconvenient. However, the absorption spectrum will show that a frequency doubled Nd^{3+} laser makes an almost ideal pump source for $Ti^{3+}:Al_2O_3$.

Our $Ti^{3+}:Al_2O_3$ sample[16] was a rod 1 inch-long, 1/4 inch in diameter, and 60° Z cut, with the ends polished and broad band AR coated around 7800 Å. The amount of TiO_2 in the sample was measured by x-ray

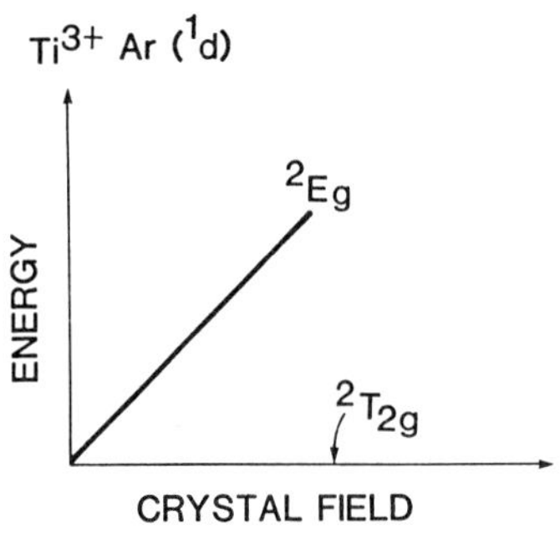

Fig. 1 Tanabe Sugano diagram for the single α electron of Ti^{3+} in a strictly octahedral environment. Slight symmetry perturbations will split the upper and lower level according to their degeneracy.

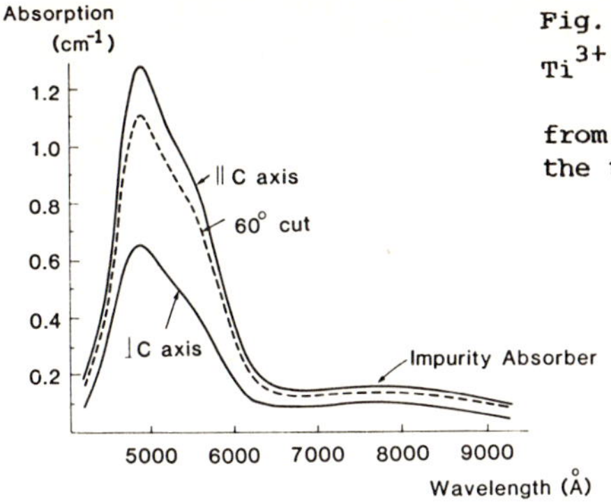

Fig. 2. Intrinsic absorption of $Ti^{3+}:Al_2O_3$. The background absorption from 7000 Å to 9000 Å is discussed in the text.

fluorescence to be 0.07 ± 0.01% by weight, which is about 7 percent of the amount that was added to the melt. We did not try to measure the relative amounts of the various Ti ionic species.

Figure 2 shows the intrinsic absorption of our sample as a function of wavelength. The measurements were made in a Cary spectrometer for light polarized perpendicular to the C axis and parallel to the C axis but incident at 60°, as this is the way the rod was cut. The third curve in Figure 2, labeled ∥C, can then be deduced via

$$A\left[C_{60°\mathrm{cut}}\right] = A\left[C_{\parallel}\right]\sin^2 60° + A\left[C_{\perp}\right]\cos^2 60°$$

where A stands for absorption. For the derivation of this expression see Réf. 20.

The main feature in the absorption spectrum is a broad peak centered around 5000 Å, which corresponds to the $^2T_{2g} \rightarrow {}^2E_g$ transition broadened by lattice vibrations. The shoulder on the long wavelength side is a splitting of the ^2E_g state due to the Jahn Teller effect. Since the Ti^{3+} environment in Al_2O_3 is not precisely octahedral but is perturbed by lower symmetries, the degeneracies of upper and lower state will be lifted and additional splitting will take place. To resolve these effects, low - temperature absorption measurements would have to be made.

The broad absorption feature from 6500 Å to 9000 Å is of unknown origin, although it has meanwhile been shown[17] that its appearance depends on details of the crystal growth process as well as the doping concentration. Samples with considerably reduced absorption in this wavelength regime have been grown[17]. Further pursuit of these efforts is essential. An absorption of ~10%/cm over the tunable lasing wavelength is clearly undesirable, since a high gain-to-loss ratio must be utilized to maintain efficient extraction.

Figure 3 shows the fluorescent emission, corrected for the S1 photomultipler and the spectrometer response. The errorbar is a representative estimated error, largely resulting from an uncertainty in the temperature of the Tungsten lamp used to calibrate the spectrometer and a strong decrease in grating efficiency towards 1 μm wavelength. The

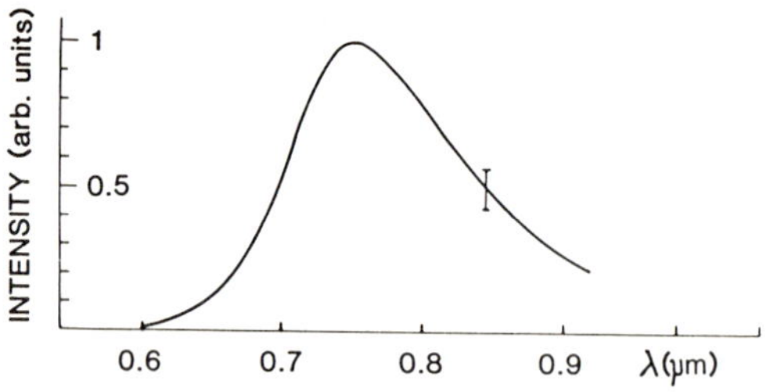

Fig. 3 Fluorescent emission of Ti^{3+} in Al_2O_3. The errorbar is a representative estimated error due to calibration uncertainties.

sample was pumped with a frequency doubled YAG laser. The fluorescent emission lineshape is independent of the orientation of polarization of the fluorescence and the pump as well as the crystal orientation. The strength of the emission, however, varies as a function of fluorescence polarization angle, as shown in Figure 4. Dots are measurements on the 60° Z cut rod, and the data was found to fit the theoretical curve of $p = 0.22 \cos^2\theta + 0.78$, as indicated. For a 90° cut rod, the polarization dependence of the fluorescence is projected to be $p = 0.26 \cos^2\theta + 0.82$. Assuming no nonradiative transitions, this data, together with the 3.2 μsec fluorescent lifetime, can be reduced to yield a stimulated emission cross-section at line center of

$$\sigma_p = (3 \pm 0.5) \times 10^{-19}\ \mathrm{cm}^2 \parallel \mathrm{C\ axis}$$

$$\sigma_p = (1.4 \pm 0.3) \times 10^{-19}\ \mathrm{cm}^2 \perp \mathrm{C\ axis}$$

which implies a minimum homogeneous saturation energy density of ~0.97 J/cm^2. Even though other investigators report somewhat smaller cross-section values[18], these numbers are a good order of magnitude above those of e.g., Alexandrite.

We also operated a very simple $Ti:Al_2O_3$ oscillator. The rod was placed in a cavity about 5 cm long with a R = 100%, 10 cm radius of curvature mirror with R = 95% and R = 75% flat output couplers. The sample was pumped longitudinally through the curved mirror in such a way that the

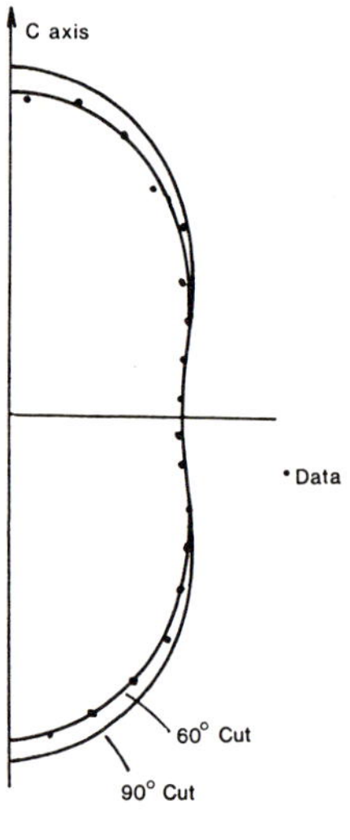

Fig. 4 Polar diagram of the emission strength as a function of angle with respect to the C-axis for a 60° cut (measured) and a 90° cut (predicted).

pump light was collimated through the sample with a ~1.2 mm-diameter beam. The source was a flashlamp-pumped dye laser operating at 0.504 μm. The sample absorption at the lasing wavelength, as discussed above, is responsible for the almost indistinguishable thresholds when the mirror reflectivity is changed from 90 percent to 75 percent (Figure 5). However, lower reflectivity output couplers would result in higher slope efficiencies and eventually separate the thresholds.

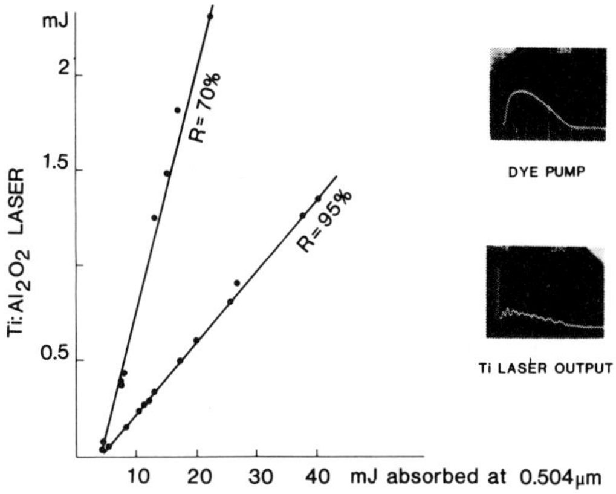

Fig. 5 Slope efficiency measurements for R = 70% and R = 95% output couplers. The inserts show the dye laser pump pulse and the resulting $TiAl_2O_3$ laser output.

Since these measurements do not allow an estimate of the transfer efficiency from pump photons to inverted states in the usual way, one can estimate the transfer efficiency as follows.

Since the sample loss is known (Figure 2), as are the reflectivity of the output coupler and the simulated emission cross-section, the number of inverted states required to reach threshold is easily calculated. Measuring the number of pump photons from the dye laser to achieve threshold in the $Ti:Al_2O_3$ laser then shows that the above transfer efficiency is ~1.

Even though $Ti:Al_2O_3$ still needs some development at this point, its advantages of no upper state absorption, of using a host material whose growth technology has advanced towards large boule sizes[19] with good optical quality, and of a large stimulated emission cross-section make it a desirable alternative to Cr^{3+} doped systems for efficient short pulse operation.

Acknowledgements

We thank P. Moulton for valuable and instructive discussions and Ed Mathez for performing the x-ray fluorescence measurements.

REFERENCES

1. P.F. Moulton: Appl. Phys. Lett. 35, 838 (1979)

2. J.C. Walling, H.P. Jenssen, R.C. Morris, E.W. O'Dell, and O.G. Peterson: Opt. Lett. 4, 182 (1979)

3. D.J. Erlich, P.F. Moulton, and R.M. Osgood: Opt. Lett. 4, 184 (1979).

4. P.F. Moulton, R.E. Fahey, and W.F. Krupke: LLL Annual Report UCRL-50021-80, Vol. 3, (1980) pp. 8-60

5. D.J. Erlich, P.F. Moulton, and R.M. Osgood: Opt. Lett. 5, 339 (1980)

6. M.L. Shand, an J.C. Walling: IEEE JQE 18, 1829 (1982)

7. B. Struve, and G. Huber: "Laser Action and Broad Band Fluorescence in Cr^{3+} Ga Sc Gd - Garnet," Proceedings of XIIth International Quantum Electronics Conference, 1982, p. 235

8. P. Moulton, "Ti-Doped Sapphire: A Tunable Solid-State Laser," Optics News, Nov./Dec. 1982, p. 9

9. L.F. Johnson, H.J. Guggenheim, D. Bahnck, and A.H. Johnson: Opt. Lett. 8, 371 (1983)

10. D.E. McCumber: Phys Rev. 134, A299 (1964)

11. L.F. Johnson, H.J. Guggenheim, and R.A. Thomas: Phys. Rev. 149, 179 (1966)

12. P. Moulton: "New Developments in Solid State Lasers," Laser Focus, May 1983, p. 83

13. L. De Shazer in Laser Technology Group Meeting, P. Moulton, Modertor, OSA Annual Meeting, New Orleans, 1983

14. P.T. Kenyon, L. Andrews, B. McCollum, and A. Lempicki: IEEE JQE 18, 1189 (1982)

15. A. Budgor, "Overview over Chromium Doped Tunable Vibronic Lasers," Proc. SPIE, Vol. 461, (1984) p. 64

16. The sample was supplied by Union Carbide Corporation, San Diego, California

17. private communication from Union Carbide Corp.

18. P. Moulton, see Reference 19. $\sim 2 \times 10^{-19}$ cm^2 for ∥C axis were reported

19. F. Schmidt, and C.P. Khattak: "Large Crystal Sapphire Optics," Laser Focus, p. 147, (September 1983)

20. Born and Wolf, 5th Ed., p. 708-711, Pergamon Press 1975

Stimulated Emission from Flashpumped $Ti:Al_2O_3$

L. Esterowitz, R. Allen

Naval Research Laboratory, Washington, DC 20375, USA

C.P. Khattak

Crystal Systems, Inc., Salem, MA 01970, USA

Stimulated emission from $Ti:A\ell_2O_3$ was first reported by P. MOULTON [1] in 1983 using laser pumping. Further improvements and an increased tuning range were recently reported by P. MOULTON [2] at CLEO '84, again using laser pumping.

In this work we describe the first operation of trivalent titanium in $A\ell_2O_3$ (sapphire) as a laser under flashpumped excitation. The $Ti:A\ell_2O_3$ was grown by Crystal Systems, Inc. with a titanium concentration of nominally 0.1% by weight. The rod was cored along the growth axis of the 2.5 inch diameter boule so that the c axis was perpendicular to the rod axis. The dimensions of the rod after polishing were 7 mm diameter x 49 mm.

The crystal had appreciable scattering centers and had gradation of Ti concentration along the length of the rod.

The trivalent titanium ion has only two d-electron energy levels in a crystalline environment. Since there are no d-state levels above the upper laser level the negative effect of possible excited-state absorption is avoided. The stimulated emission cross-section as measured by P. MOULTON [2] and G. F. ALBRECHT [3] was 2 to 3 x 10^{-19} cm^2. The fluorescent lifetime of 3.2 μsec measured by us agreed with that of P. Moulton.

The $Ti:A\ell_2O_3$ rod was pumped coaxially by a xenon flashlamp. The coaxial configuration consisted of two concentric quartz tubes with electrodes, with the laser rod inserted in the interior of the lamp. The space between the laser rod and the inside wall of the lamp was used for a circulating liquid coolant. The power supply electronics and cabling was adjusted so that the flashlamp pulse width was 5 μsec.

Laser action was first observed using water coolant at a threshold of 34J. Since most of the flashlamp energy is emitted at wavelengths much shorter than the titanium absorption bands a coumarin dye was added to the coolant to act as a fluorescent converter. The coumarin 480 dye converted an appreciable portion of the ultraviolet flashlamp emission into a region overlapping the titanium absorption spectrum. This decreased the threshold for laser action by a factor of five and increased the laser output energy by more than an order of magnitude.

The laser cavity mirrors used were 40 cm apart with a 96% partial reflector and a total reflector in a flat-flat configuration. The titanium laser rod was anti-reflection coated, with the 3% points at 660 and 900 nm. A summary of the lasing conditions are given in Table I.

The titanium laser was prism tuned between 710 and 912 nm. The $Ti:A\ell_2O_3$ fluorescence together with the measured laser tuning range are shown in Fig. 1. The laser cavity mirror reflectivities decrease sharply

Table 1. Lasing Conditions of $Ti^{3+}:A\ell_2O_3$

Dimensions of Rod	7mm x 40 mm active length
Concentration	0.1% Ti^{3+}
Pumping Method	Coaxial Flashlamp
Mirrors	96% Partial Reflector, Flat-Flat Configuration
Tuning Range	710-912 nm (limited by mirrors and AR coating)
Cavity Length	40 cm
Ti^{3+} Lifetime	3.2 μsec
Flashlamp Pump Width	5 μsec
AR Coating	Min. Reflectivity .25% 3% points 660 and 900 nm

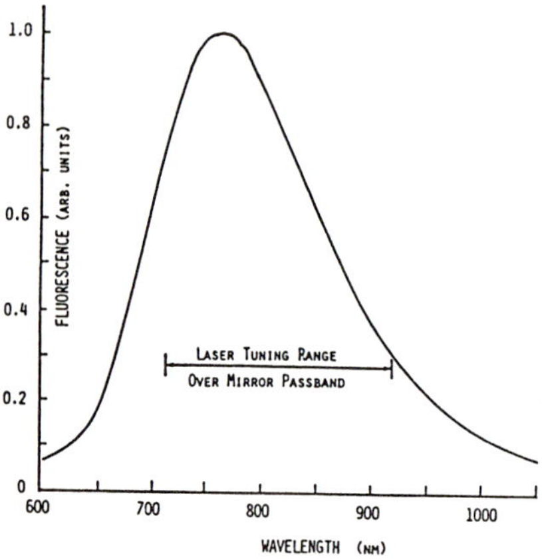

Fig. 1 Titanium fluorescence

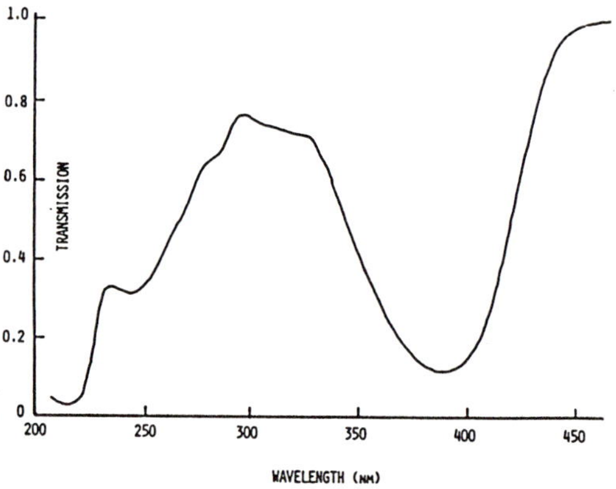

Fig. 2 Coumarin dye absorption

at 710 and 912 nm so that the tuning range was primarily limited by the mirror passband

The absorption (or transmission) spectrum of the C480 coumarin dye that was added to the coolant is shown in Fig. 2. The effectiveness of the C480 dye as a fluorescent converter is depicted in Fig. 3 by noting the excellent overlap between the C480 fluorescence and the $Ti:A\ell_2O_3$ absorption.

A plot of laser energy out versus flashlamp energy in is given in Fig. 4.

Efforts to increase the efficiency of the $Ti:A\ell_2O_3$ laser are currently being pursued. An improved pumping configuration with a better fluorescent converter together with a higher quality single crystal material should significantly affect ultimate $Ti:A\ell_2O_3$ flashpumped laser performance.

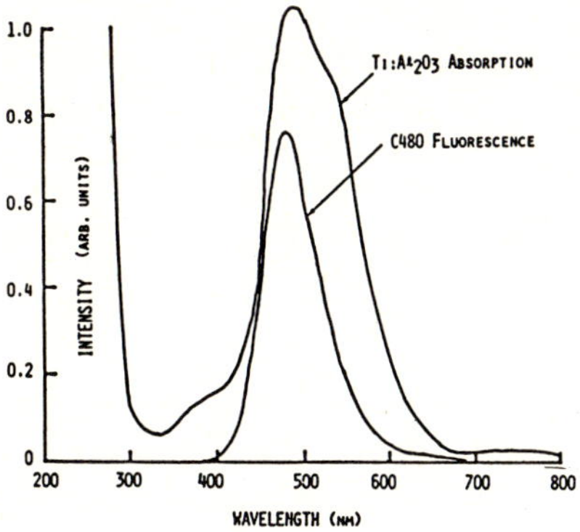

Fig. 3. Titanium absorption and C480 fluorescence

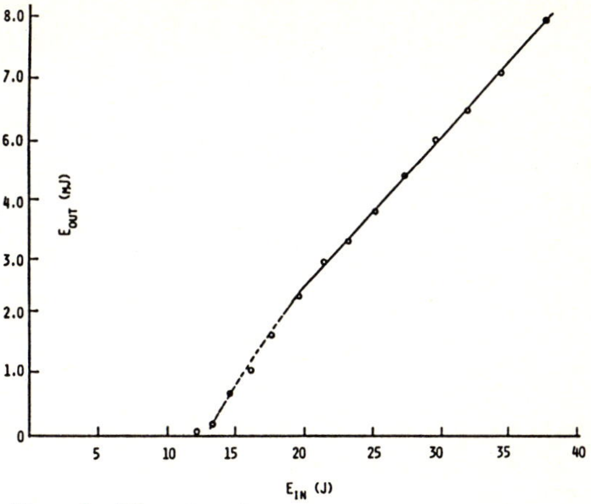

Fig. 4 Titanium laser energy out as a function of flash-lamp energy

[1]P. F. Moulton, Laser Focus, p. 83, May 1983.

[2]P. F. Moulton, CLEO Tech. Digest, WA2, p. 77, June 1984.

[3]G. F. Albrecht, J. M. Eggleston and J. J. Ewing, 1st Annual Conference on Tunable Solid State Lasers, June 1984, La Jolla Inst., La Jolla, CA.

Laser Pumped Single Pass Gain

M.L. Shand and Shui T. Lai
Applied Corporation, Corporate R&D, Morristown, NJ 07960, USA

Laser pumped single pass gain measurements are particularly useful for small samples and are sufficiently accurate to determine excited state absorption cross-sections in solid state tunable laser materials. Preliminary results for alexandrite and emerald are presented.

1. Introduction

Flashlamp pumped single pass gain (SPG) measurements have been used to determine the excited state absorption cross-section, σ_{2a}, of Cr^{3+} throughout the vibronic lasing region of alexandrite [1]. The wavelength dependence of σ_{2a} together with the emission cross-section, σ_e, and the ground state absorption cross-section, σ_a, have been used to model alexandrite laser performance [2]. Two problems arise in applying this technique to new potential laser materials. First, the sample must be available in sufficient size to make rods of at least one to two inches long. Second, a method for determining the excited state ion density must be found. For the alexandrite measurements, the decrease in the sharp R line ground state absorption was used to determine the percentage of Cr^{3+} ions in the excited state. Many of the Cr^{3+} laser hosts discussed in these proceedings [3] do not have measurable R line absorption.

The laser pumped SPG measurement overcomes both these problems. First, the sample length needed is only a few mm because of the strong and localized absorption of the pump light. Second, the number of excited state ions can be calculated from the measured absorbed pump light. These features of the laser pumped SPG measurement will be demonstrated in the experiments described below. The results are preliminary, however, they show that σ_{2a} can be extracted from these experiments.

2. Experiment and Analysis

A 647 nm Kr laser line pumps both the sample and a dye laser, which provides a probe beam. The pump beam is chopped, typically on for 1-5 ms at 10-20 Hz, to reduce heating in the optics and in the sample. The probe beam is coincident with the pump beam, with the overlap monitored by observing the thermal lensing experienced by the probe beam. The pump and probe beam are focussed at the sample position. The probe beam power has a dc component representing the power of the transmitted beam through the unpumped sample and an ac component representing the power increase or decrease (gain or loss) of the transmitted beam through the pumped sample. The ratio of probe beam power with pumping to probe beam power with no pumping, P_p/P_u, is given by $1+P_{ac}/P_{dc}$. The probe beam intensity is kept sufficiently low to avoid gain saturation. The maximum gain is less than 2% so that parasitic losses such as amplified spontaneous emission can be ignored.

The excited state ion density can be calculated from the absorbed pump power. Excited state absorption consumes part of the pump photons. A correction for the pump beam excited state absorption is made by measuring the SPG as a function of pump power. At low powers the SPG is proportional to pump power. At high powers, the SPG is less than linear with pump power, as part of the pump light contributes to excited state absorption. Extrapolation of the linear portion of SPG vs pump power gives the expected SPG if no excited state absorption of the pump were present. This expected value of SPG is used in the analysis of the data.

The analysis is based on the fact that both the probe and pump beams and therefore the excited state ion density, $N_2(r)$, have Gaussian spatial profiles [4]. The probe beam intensity after the sample for the two cases of the sample pumped, $I_p(r)$, and of the sample unpumped, $I_u(r)$ are given by

$$I_p(r) = I_o(r)\exp[N_2(r)(\sigma_e-\sigma_{2a}+\sigma_a)\ell-N\sigma_a\ell-L] \quad (1)$$

$$I_u(r) = I_o(r)\exp[-N\sigma_a\ell-L] \quad (2)$$

where ℓ is the sample length, N is the total Cr^{3+} ion density, and L accounts for the scattering losses in the sample and on all optical surfaces.

The exponential factor in (1) is generally small, thus the exponential function can be expanded in a power series. Taking the series to second order and integrating (1) and (2) over the beam area gives the power of the probe beam in the two cases. The resulting ratio is

$$P_p/P_u = 1 + \frac{2(\sigma_e-\sigma_{2a}+\sigma_a)\alpha P_{pump}\tau}{h\nu_{pump}\pi(\omega^2_{pump} + \omega^2_{probe})} + \frac{2(\sigma_e-\sigma_{2a}+\sigma_a)^2\ (\alpha P_{pump}\tau)^2}{(h\nu_{pump}\pi\omega_{pump})^2(\omega^2_{pump}+2\omega^2_{probe})} \quad . \quad (3)$$

Note that this expression depends neither on sample length nor on Cr^{3+} concentration in the sample.

The value of σ_{2a} is obtained from (3) as all the other parameters can be either measured or calculated. P_p/P_u is the SPG result, σ_e and σ_a are determined from fluorescence measurements and the extended McCumber theory [1], and the beam waists are determined from the beam divergence which is measured with a diode array.

3. Results and Discussion

The apparatus has been tested with alexandrite, for which σ_{2a} is known [1]. The calculated σ_{2a} are shown in Fig. 1. The probe beam was polarized parallel to the b axis of alexandrite, which is the only polarization direction showing gain. The values of σ_a are very close to those measured previously with flashlamp pumping.

The results for emerald are shown in Fig. 2 for polarizations parallel and perpendicular to the hexagonal axis (c axis). The ratio of σ_{2a} to σ_e for emerald in the gain region for E//c is comparable to that of alexandrite. The σ_{2a} for both polarizations are comparable, so that for E⊥c, σ_{2a} is a significant fraction of σ_e. This result implies that the E⊥c direction may not be as useful for lasers as previously expected.

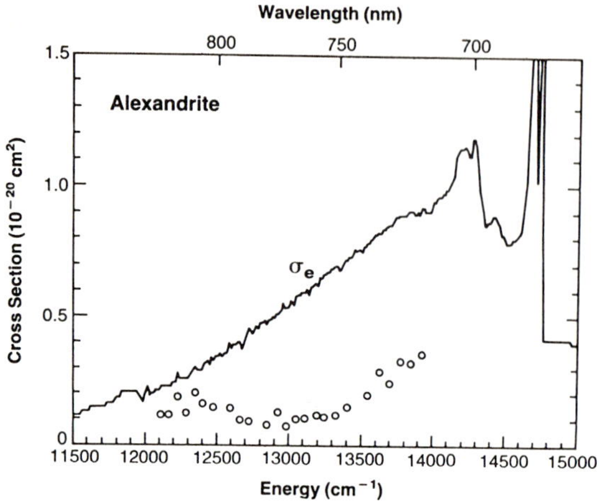

Fig. 1 Emission cross-section (solid-line) and excited state absorption cross-section (circles) for alexandrite E//b

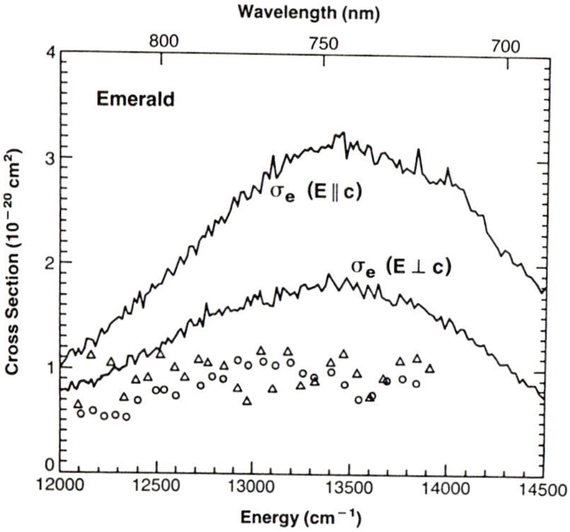

Fig. 2 Emission cross-section for emerald, polarization E//c and E⊥c (solid lines); excited state absorption cross-section E//c (circles) and E⊥c (triangles)

The major source of error in determination of σ_{2a} is in determining the beam waists. The main difficulty is that the Kr laser takes a long time (>35 hours) to reach thermal equilibrium. During this time the pump beam waist changes by over 25%. The probe beam waist is relatively stable and is also smaller than the pump beam waist by a factor of three to limit thermal lensing effects, so that the pump beam waist error dominates the error in σ_{2a}. Efforts are now underway to improve the pump beam waist stability. Until these efforts are complete, these results must be considered preliminary.

5. Conclusion

The laser pumped SPG measurement can be used to determine σ_{2a}. The system has been tested with alexandrite and used to determine σ_{2a} in emerald. The major problem at the moment is stability of the pump source.

6. Acknowledgements

We thank R.C. Morris and H.P. Jenssen for helpful discussions. This project is supported in part by the U.S. Army Research Office, Research Triangle Park, NC.

References

1. M.L. Shand and J.C. Walling, IEEE J. Quantum Electron. QE-18, 1152, 1982; M.L. Stand and H.P. Jenssen, IEEE J. Quantum Electron. QE-19, 480, 1983.

2. R.C. Sam, these proceedings; D.F. Heller and J.C. Walling, Conference on Lasers and Electro-Optics 1984, Opt. Soc. of Amer., p. 102.

3. G. Huber and K. Petermann, these proceedings.

4. S.T. Lai and M.L. Shand, J. Appl. Phys. 54, 5642 (1983).

Trivalent Cerium Doped Crystals as Tunable Laser Systems: Two Bad Apples

Douglas S. Hamilton

Department of Physics and Institute of Materials Science, University of Connecticut, Storrs, CT 06268, USA

The 5d→4f transitions of trivalent doped crystals have broad emission bands with large oscillator strengths and near unity quantum efficiency. These characteristics make cerium systems strong candidates for tunable solid state lasers. However, two such cerium crystals will probably never lase. The first is Ce^{3+}:YAG where a strong excited state absorption quenches the lasing transition. Our recent measurements have indicated that the excited state absorption terminates in the YAG conduction band with a peak cross-section of 1.0×10^{-17} cm^2 at 700 nm. Some of the general features of impurity ion to band spectra are discussed. The second system is Ce^{3+}:CaF_2 where a UV pump-induced photochromic center is produced following excitation of the cerium ions. The initial measurements of cerium-related transient absorptions in Ce^{3+}:YLF are also presented.

1. Introduction

Optical transitions involving the Lanthanide ions as doped constituents of crystals fall into two distinct groups, the 4f→4f' and the 4f→5d transitions. One important difference between these two groups is that the 4f→5d transitions are electric dipole-allowed whereas the 4f→4f' transitions are electric dipole-allowed only to the extent by which odd crystal field components have mixed opposite parity orbitals into the 4f states. Thus the 4f→4f' transitions have small electric dipole matrix elements ($\mu \sim 10^{-20}$ esu) and long radiative lifetimes ($T \sim 10^{-4}$ sec) compared to those for the 4f→5d transitions, for which $\mu \sim 10^{-18}$ esu and $T \sim 10^{-8}$ sec.

A second characteristic difference between the two types of transitions is in the strength of the electron-lattice coupling. The origin of this difference can be understood from Fig. 1. An electron in a 4f orbital is well shielded from the crystal fields

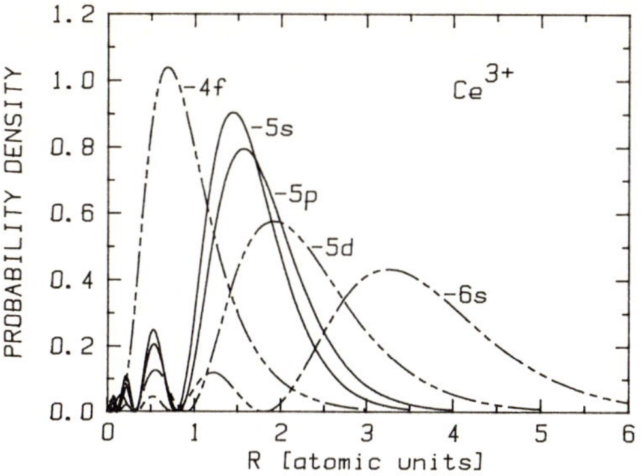

Fig. 1. The radial probability density for various outer electron orbitals in trivalent cerium-calculated using a multiconfiguration Hartree-Fock program [1]

by the electrons which fill the 5s and 5p orbitals. For an electron occupying a 5d orbital, this shielding is less complete. It is the difference in the electron-lattice coupling between the two states which participate in an optical transition which determines the linewidth and shape. Since the difference in V_{el} between two 4f states is typically very small, the corresponding optical transitions are sharp no-phonon lines with weak phonon sidebands. The larger difference in V_{el} between 4f and 5d orbitals is responsible for the large vibronic linewidths observed for the 4f→5d transitions.

Of the various lanthanide ions, the energy level diagram of trivalent cerium is the simplest since Ce^{3+} has only a single optically active electron. The spin orbit interaction splits the 4f level into the $^2F_{5/2}$ and $^2F_{7/2}$ states separated by typically 2,000 cm^{-1} and the crystal field interactions are responsible for additional splittings within each manifold of order 100 cm^{-1}. For the 5d levels, a stronger electron lattice-coupling produces typically five (for low symmetry Ce^{3+} sites) crystal field split states with energy separations of order 5,000 cm^{-1}. The room temperature absorption and fluorescence spectra of Ce^{3+}:YAG, shown in Fig. 1, reflect the characteristic features of the energy level stucture. Because of the large energy gap between the lowest 5d level and the 4f levels (typically 20,000 to 30,000 cm^{-1}) the probability for multiphonon mediated nonradiative relaxation is small and the 5d→4f quantum efficiency of fluorescence is nearly unity.

The large 5d→4f oscillator strength, broad vibronic linewidth and unity quantum efficiency are the characteristics of Ce^{3+}

which have made it an attractive ion for tunable solid state laser applications. In addition, such cerium lasers would operate at visible and near ultraviolet wavelengths (depending on the host) which is a spectral interval outside the typical operating region for transition metal lasers. However, only two Ce^{3+} crystals are known to show gain and laser action, Ce^{3+}:YLF [3] and Ce^{3+}:LaF_3 [4]. In this article, we will review our work on two Ce^{3+} systems which will probably never lase (two bad apples in the cerium bushel). In the first material, Ce^{3+}:YAG, it is a strong excited state absorption (ESA) which quenches the optical gain at the fluorescence wavelengths [5-8]. For Ce^{3+}:CaF_2, the 308 nm resonant pumping of the lowest 5d level results in the production of long-lived photochromic centers [9,10]. We will also present our initial measurements of a transient absorption process in Ce^{3+}:YLF which may be responsible for its less than optimal performance as a laser material.

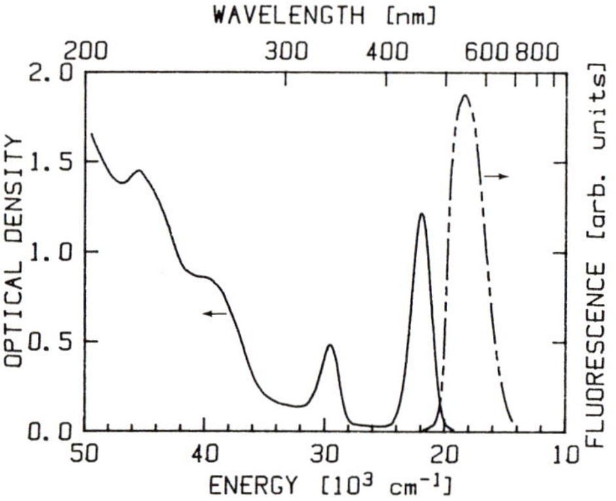

Fig. 2. The absorption and fluorescence spectra of 0.1% Ce^{3+}:YAG sample 0.5 cm in length. The onset of the strong absorbtion at 200 nm reflects the 50,000 cm^{-1} bandgap of the YAG host [2]

2. Excited State Absorption in Ce^{3+}:YAG

The excited state absorption (ESA) process is illustrated schematically in Fig. 3. Unlike two-photon absorption, ESA involves a real intermediate state which is populated via the ground state absorption (GSA). In a vibronic system, there will in general be a subsequent "cooling" transition involving accepting mode phonons before the absorption of the second photon. In laser applications, the photons involved in the ESA can either be those employed in the optical pumping cycle or those emitted by other ions during a radiative relaxation. In

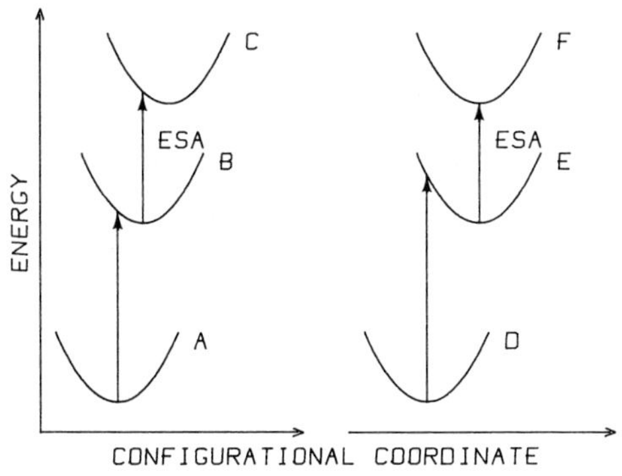

Fig. 3. Although the ground state absorption spectra for both cases consist of broad vibronic bands, the ESA transition on the right is a sharp no-phonon line because of the small difference in the electron-lattice coupling strength between the two excited states E and F

either case, the ESA reduces the net gain and raises the laser threshold. In addition, if the terminal level of the ESA relaxes to the lower states by phonon emission, an extra thermal load is placed on the system.

There have been three previous investigations of the strong ESA transition in Ce^{3+}:YAG by pumping the second 5d state, which populated the lowest 5d state through an intraconfigurational relaxation, and then measuring the transmission of a probe beam over the wavelength region of the fluorescence emission [5-7]. The ESA cross-section was seen to increase monotonically with increasing probe wavelength and did not seem to match the edge of any spectral features known from the ground state absorption spectrum.

We have reinvestigated [8] the ESA in this material using the experimental setup illustrated in Fig. 4. The excitation source

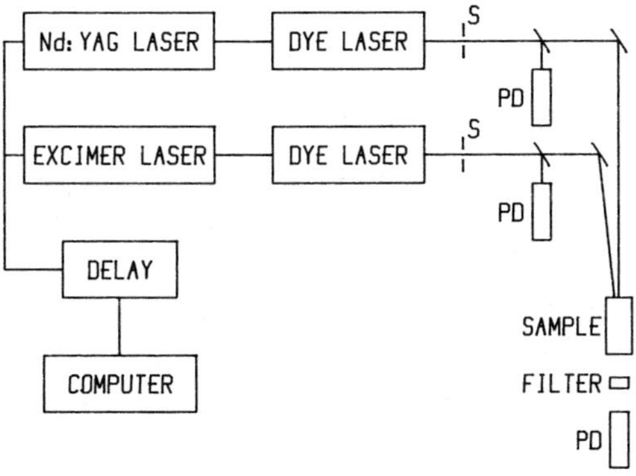

Fig. 4. For the ESA measurements, the outputs of the three calibrated photodiodes are read by a computer interface to monitor the probe transmission and the pump intensity

for the GSA was an excimer laser-pumped dye laser tuned to a 500 nm wavelength. At this wavelength, the excitation is into the low energy tail of the lowest 5d absorption band where $\alpha L \sim 1$. This results in a lower thermal excitation of the sample compared to that from pumping into the second 5d band. Moreover, since color centers in YAG can be created for excitation wavelengths shorter than 400 nm [11], this longer wavelength pumping will eliminate any pump-induced transient centers. A Nd:YAG laser-pumped tunable dye laser was used to generate the weak probe beam whose transmission through the sample is a measure of the ESA. A computer controlled time delay generator was used to set the pump to probe delay in 10 nsec increments. The typical values of the beam diameters and pulse energies at the sample were 1 mm and 0.1 mJ for the pump and 0.4 mm and 5 nJ for the probe.

For an ESA originating on a single state, there will be a simple exponential dependence of the ESA on the pump to probe delay. The measurements of the decay with 500 nm pumping has a single 65 nsec decay constant which is the radiative lifetime of the lowest 5d state, thus demonstrating that this is then the initial state of the ESA transition. The measured ESA increases linearly with increasing pump flux, and thus there are no complications from multi-photon or saturation effects.

The spectral dependence of the ESA is displayed in Fig. 5. The characteristic tail on the high-energy side is strikingly similar to that in the absorption spectrum of the F' center in KBr which is a photoionization transition to the KBr conduction band [12]. We have also assigned the terminal state of the ESA transition in Ce^{3+}:YAG as the YAG conduction band which is

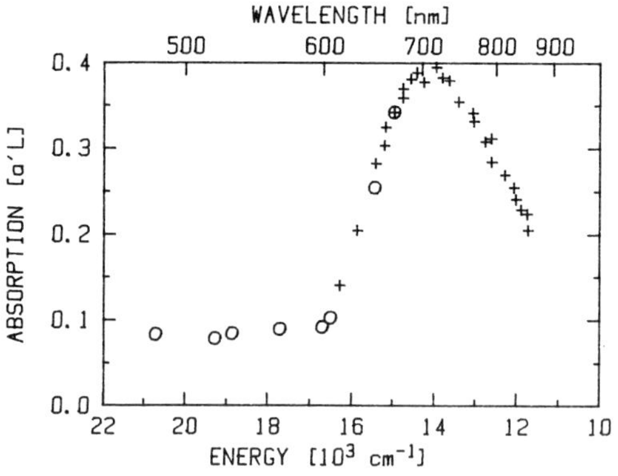

Fig. 5. The ESA spectrum for Ce^{3+}:YAG at 300 K. The open circles are values from Miniscalco et al. [6] and the crosses are from Gayen et al. [8]

supported by the large value of the ESA cross-section ($\sigma' = 1.0 \times 10^{-17}$ cm^2 at 700 nm). Such a large value is inconsistent for a 5d→5d' ESA but is typical of photoionization cross-sections measured in both insulators [12] and semiconductors [13].

We can overlay the ESA and the GSA by adding the energy of the ground state to lowest 5d state no-phonon transition [14] to the ESA spectrum. This superposition is illustrated in Fig. 6 and does not show any correlated structures between the two spectra. It is important to note that when localized-to-band transitions are involved, one can not simply add energies together and use the GSA to predict the ESA. A localized-to-band transition will peak at an energy, relative to a band threshold, which depends on the characteristic spatial extent of the localized state [13]. More diffuse states (such as a 5d state) will have their maximum transition strength at a lower relative energy than will more compact states (such as a 4f state). These comments also suggest that the rising background in the GSA is the 4f to conduction band transition. We will say more about this later.

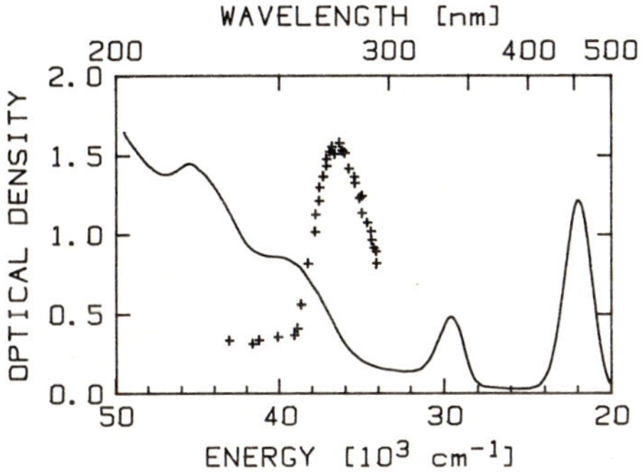

Fig. 6. An overlay of the GSA spectrum and the ESA spectrum relative to the lowest 5d level

The assignment of the states which participate in an optical transition from just spectral information is a bit risky however. Other characteristics of the participating states, such as the radiative lifetime or quantum efficiency of fluorescence are always helpful. To further clarify the nature of the terminal state of the ESA transition in Ce^{3+}:YAG, we have measured the relaxation from the terminal state back to the lowest 5d state. The idea here is to use an intense probe pulse which can carry as many electrons as possible into the terminal state of the ESA transition and then to simultaneously monitor the 5d→4f fluorescence which is proportional to the instantaneous population of the 5d state.

This strong probe saturation experiment was conducted with a 50 nJ energy and 1.8 mm diameter pump pulse at 500 nm and a concentric probe pulse at 700 nm with a 1 mJ energy and a 2.5 mm diameter. The resulting oscilloscope trace of the 5d→4f fluorescence for a 20 nsec pump to probe delay is illustrated in Fig. 7. Following the pump pulse, the fluorescence begins to decay with its 65 nsec radiative lifetime. The probe pulse almost totally depopulates the 5d level and the fluorescence signal drops by 70%. Most important is the fact that the fluorescence signal does not return to its preprobe value, that is, the 5d population does not seem to rapidly refill from the terminal state of the ESA. This demonstrates that there is not a rapid relaxation from the terminal state back to the lowest 5d state which seems to rule out the higher 5d configurations as the terminal state.

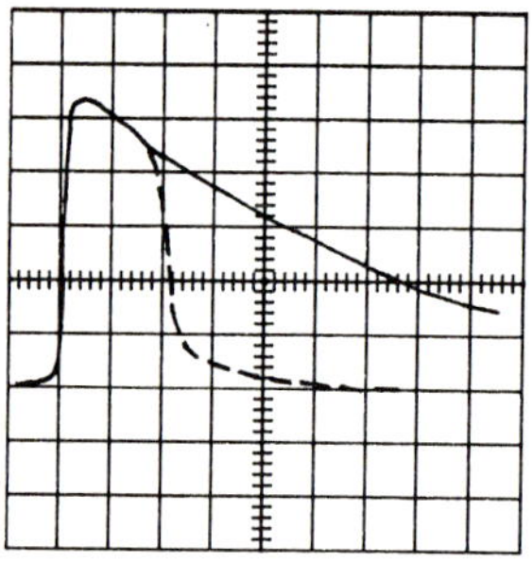

Fig. 7. The oscilloscope trace (10 nsec/div and 10 mV/div) shows the pump-only fluorescence with a solid curve and the pump with probe fluorescence with a dashed curve

Additionally, there is a weak afterglow which is observed in the presence of the strong probe pulse. This afterglow is not observed when the probe pulse is blocked and drops in intensity with increasing pump to probe delay. This µsec tail has a lifetime which is characteristic of the afterglow observed in Ce^{3+}:YAG with cathode ray excitation and which reflects the carrier mobility in the YAG host [15]. A measurement of the time-integrated fluorescence of the lowest 5d state shows that the refilling of the 5d level is quite inefficient, however. By replacing the oscilloscope by a Keithly electrometer, we observe a maximum 60% drop in the time-integrated fluorescence at a 5 nsec pump to probe delay. This suggests that significant recombination is occuring through pathways other than through the upper cerium levels.

As mentioned previously, we have interpreted the rising background in the GSA spectrum as the 4f to conduction band transition. Both this rising background and the ESA spectrum in Fig. 6 tend toward zero at about 30,000 cm^{-1}, which we take as the 4f to conduction band threshold. If the sample is pumped at 308 nm, which is in the valley between the second and third 5d resonances, then the ESA developes a long μsec tail in addition to the short 65 nsec component. This tail reflects the electrons which have been promoted to metastable trap states via a photo-ionization of the cerium ions. A weak afterglow is also observed after 308 nm excitation which is not obseved with 500 nm pumping.

3. Ultraviolet Induced Photochromic Centers in $Ce^{3+}:CaF_2$

Recently, we have completed a study of two-photon transitions to the lowest 5d state of the cerium ions in CaF_2 [16,17]. The analysis of this nonlinear process would benefit from a knowledge of the dipole matrix elements between the various 5d levels which could be obtained from ESA measurements. These transient absorptions [9] were partially obscured by a large and persistent background absorption induced by the resonant pumping of the lowest 5d level of the cerium ions with a 308 nm XeCl excimer laser. The crystals, which where originally transparent before exposure, turned a dirty brown after the UV irradiation. Off resonant pumping, either at 337 nm or 351 nm did not color the cerium doped samples, nor was any coloration observed in undoped CaF_2 crystals [10]. Thus the centers are created in a process involving the the cerium ions in an excited configuration.

The absorption spectrum of the samples after the 308 nm exposure shows three overlapping bands with peaks at 390 nm, 510 nm and 800 nm. The sample returns to its initial transparent state after exposure to 351 nm, 532 nm or 633 nm light. The fact that bleaching of these photochromic centers occurs at wavelengths corresponding to the 5d→4f fluorescence of the Ce^{3+} ions, complicates the rate equation for N, the density of these centers, by introducing a cross-term proportional to Nn', where n' is the excited cerium density. However, if we solve the rate equation for N after the first pulse in an crystal, then all of the N→n'

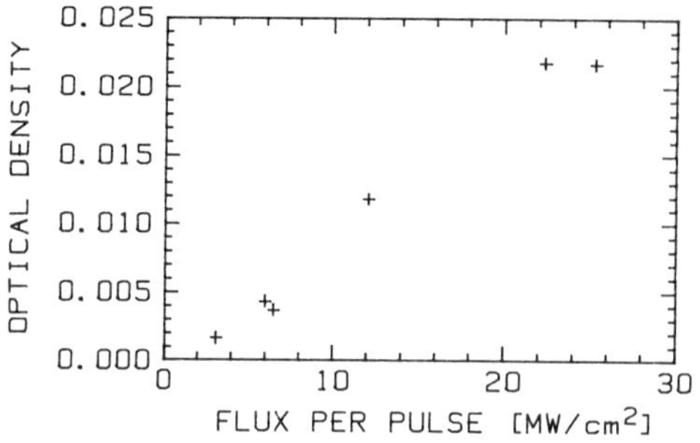

Fig 8. The absorption of 633 nm light after a single first 308 nm optical pulse as a function of the intensity of the pulse

recombination terms will be absent since N=0 before the first pulse. The measurements of the optical density following a single pulse are shown in Fig. 8. The results show a linear relationship between N and the intensity of the 308 nm pulse. Thus we infer that multiphoton processes are not involved in the creation of these centers. This suggests a production mechanism in which an excited Ce^{3+} ion can relax by either returning to its ground state directly via a fluorescence transition, or by releasing the electron in an autoionizing transition. The subsequent trapping of the electron creates the photochromic center, although a more complete description may involve a complex of the trapped electron in association with the Ce^{4+} ion.

In addition to the optically induced recombination of the trapped electron with the cerium ion, there is also a thermally activated recombination, which produces an afterglow from the sample following the 308 nm irradiation. This phosphorescence is at the wavelength of the Ce^{3+} fluorescence and thus indicates that the recombination is through the excited states of the cerium ion. There are at least two characteristic delay times for this afterglow. At room temperature, there is a long lived component (at least several hours) which is correlated with the long lived photochromic centers discussed above. In addition, the afterglow shows a shorter lifetime component with a decay time of order 10 sec. This suggests that there may be several electron trapping sites with different barrier heights for recombination with the hole trapped on the cerium site.

Although further investigations are required to develop a microscopic model for the photochromic center and the recombination dynamics, our results so far show that a cerium ion in CaF_2 which has been excited to the lowest 5d level, has a finite probablity of undergoing an autoionizing transition. This may be partially understood by referring back to the electron probability density for Ce^{3+} in Fig. 1. An electron in a 5d orbital is only partially shielded from crystal field effects and is, as well, partially shielded from the attractive Coulomb field of the cerium nucleus. This shielding is in part responsible for the binding energy of a 5d electron to the cerium nucleus. The electron release will be mediated by the lattice relaxation which follows the excitation of the electron to the a 5d orbital.

4. A Transient Absorption in Ce^{3+}:YLF

The third cerium doped crystal we want to consider here is $Ce^{3+}:LiYF_4$ (YLF). Optical gain and laser action has been demonstrated in this material on the lowest 5d→4f transition via pumping of the second 5d level with a 254 nm source [3]. There were, however, several characteristics of the measurements which indicated a less than optimal performance of the crystal as a laser material. Also noted was a strong transient loss in the detected 633 nm probe signal following 254 nm pumping, which may be related to the lasing performance.

We have just initiated a project to clarify the origin of these performance-limiting characteristics. In our initial experiments, the 308 nm output of a XeCl excimer laser was used to excite the cerium ions. This wavelength corresponds to the low energy tail of the absorption to the lowest 5d level. By pumping at this longer wavelength, the thermal energy released into the sample is much smaller than with the 254 nm pumping and there is a reduced possibility of UV induced color center formation in the YLF host. Using a collinear geometry of the pulsed 308 nm pump and a 633 nm cw probe, a 50% reduction in the detected probe beam is observed immediately following the pump pulse. By using a series of different diameter apertures between the crystal and the probe detector, we observe no dependence of the fractional loss upon the aperture characteristics or position. From this we have concluded that the loss in probe signal is not due to thermal lensing but arises from an absorptive process in the pumped volume of the crystal. The transient absorption relaxes with a 150 μsec decay time which is much longer than the 40 nsec radiative lifetime of an electron in the lowest 5d level. Thus, the initial state of the transient absorption is not the lowest 5d level but involves the electron in some other longer lived configuration.

We suspect that there is some similarity in this result to the cerium-related photochromic center produced in $Ce^{3+}:CaF_2$. The relaxation rate of the center would depend on the binding energy of the trapped electron, which may be significantly different for the two cases. Measurments of the temperature dependence of the relaxation rate and the spectral features of the transient center will help to clarify the characteristics of this center and its relationship to the Ce^{3+}:YLF laser performance.

5. Acknowledgements

I would like to acknowledge Swapan Gayen, Gerald Pogatshnik and Ki-Soo Lim, whose work towards Ph.D. degrees have produced the results I have discussed here. I would also like to note that Dr. William J. Miniscalco has collaborated with us on the Ce^{3+}:YAG work and to thank Professor Ralph Bartram for several helpful suggestions. The work has been supported by the United States Department of Energy under grant number DE-FG02-84ER45056.

References

1. C.F. Fischer: Comp. Phys. Comm. 1, 151 (1969); Comp. Phys. Comm. 4, 107 (1972).
2. G.A. Slack, D.W. Oliver, R.M. Chenko and S. Roberts: Phys. Rev. 177, 1308 (1969).
3. D.J. Ehrlich, P.F. Moulton and R.M. Osgood, Jr.: Opt. Lett. 4, 184 (1979).
4. D.J. Ehrlich, P.F. Moulton and R.M. Osgood, Jr.: Opt.Lett. 5, 339 (1980).
5. R.R. Jacobs, W.F. Krupke and M.J. Weber: Appl. Phys. Lett. 33, 410 (1978).
6. W.J. Miniscalco, J.M. Pelligrino and W.M. Yen: Appl. Phys. 49, 6109 (1978).
7. J.F. Owen, P.B. Dorain and T. Kobaysi: J. Appl. Phys. 52, 1216 (1981).
8. S.K. Gayen, G.J. Pogatshnik, D.S. Hamilton and W.J. Miniscalco: (to be published).
9. S.K. Gayen and D.S. Hamilton: J. Opt. Soc. Amer. 73, P1393 (1983).
10. G.J. Pogatshnik, S.K. Gayen and D.S. Hamilton: J. Lumin. (to be published).
11. K. Mori: Phys. Status Solidi, A42, 375 (1975).
12. R.S. Crandall and M. Mikkor: Phys. Rev. 138, A1247 (1964).
13. B.K. Ridley: Quantum processes in semiconductors (Oxford, New York 1982).
14. D.J. Robbins: J. Electrochem Soc. 126, 1550 (1979).
15. D.J. Robbins, B. Cockayne, J.L. Glasper and B. Lent: J. Electrochem. Soc. 126, 1213 (1979).
16. S.K. Gayen and D.S. Hamilton: Phys. Rev. B28, 3706 (1983).
17. S.K. Gayen, G.J. Pogatshnik and D.S. Hamilton: J. Lumin. (to be published).

Cuprous Ion Doped Crystals for Tunable Lasers

L.G. DeShazer

Hughes Research Laboratories, Malibu, CA 90265, USA

Studies of copper-doped silicate and fluoride glasses demonstrated that intense fluorescence could be obtained from cuprous ions in the 350-900 nm region, and that laser gain could be achieved under certain circumstances. However, solarization of the glass hosts by the UV pump light introduced a loss at the potential laser wavelengths, preventing laser oscillation. A crystalline host having a non-inversion symmetry site for the cuprous ions would maintain the large oscillator strength while possibly avoiding this solarization loss. We determined that Cu^{+1} doped crystals such as $RbMgF_3$ exhibit strong fluorescence apparently without the solarization problem.

1. Introduction

Since 1980 at Hughes Research Laboratories, monovalent transition-metal doped solids have been under study as a new class of materials for broadband tunable solid-state lasers. Monovalent Cu, Ag and Au ions in the periodic group IB are of particular interest since they are filled-shell ions ($3d^{10}$ ground configuration) with their lowest-energy transition $3d^9 4s$-$3d^{10}$ being parity forbidden for the free ion (see Fig. 1). When doped into solids, these ions can yield a very broad emission band with a lifetime that is not too short (~ 10 μs)

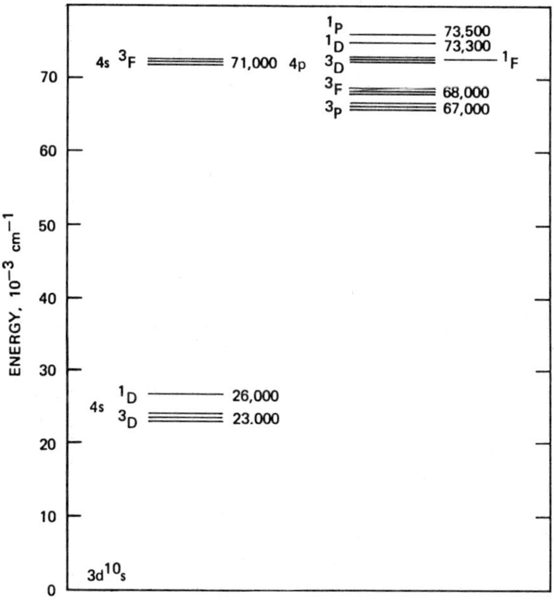

Fig. 1 Atomic energy levels of Cu^{+1} free ion [2]

for laser action. In a crystalline environment the transition becomes weakly allowed, being enforced by either lattice vibrational interactions or electronic interactions involving odd-potential crystal fields. Previously, Cu^{+1} has been doped in alkali halides [1] on sites having inversion symmetry but the emission was weak because the transition forbiddeness was broken only by lattice vibrational interactions.

When Cu^{+1} is doped into glass, however, the ions occupy sites without inversion symmetry resulting in relatively strong fluorescence from 350 to 900 nm [3]. The electrostatic field of the anions surrounding the Cu ion in glass mixes the $3d^94s$ configuration with the opposite parity configuration $3d^94p$, breaking the parity selection rule of the $3d^94s$-$3d^{10}$ transition. The transition strength is nearly 10^3 times larger than for the similarly-derived rare-earth transitions because 3d electrons are involved instead of 4f electrons, as for rare earths. In addition, divalent copper ions, which are usually present in some degree with the monovalent ions, strongly quench the monovalent fluorescence and made this broadband emission difficult to observe in past investigations. Therefore, under these considerations, our research emphasized identification of solids having monovalent copper ions on noninversion symmetry sites, and the elimination of divalent copper from the material.

2. Monovalent Copper Ions in Glass

Both oxide and fluoride glasses were prepared with Cu^{+1} concentrations near 0.3%. The lower monovalent state of Cu was obtained by producing the glasses in a reducing atmosphere. For the copper-doped silicate glass, having a base composition 0.7 SiO_2: 0.2 Na_2O: 0.1 CaO, a nitrogen/hydrogen gas mixture was bubbled through the melt. For copper-doped fluoride glasses, such as the fluorohafnate HBT composition 0.6 HfF_4:0.33 BaF_2:0.07ThF_4, the glass was prepared using a CCl_4 reactive atmosphere [4]. The fluoride glass proved to be an excellent vehicle to reduce the valency of the copper ion to its monovalent state, an improvement over the silicate glass as host. Since Cu^{+1} has no visible absorption, monovalent copper glass is water white in color. In contrast, divalent copper glass has a yellowish-green hue and, therefore, the presence of the unwanted Cu^{+2} is easily detected by material color and absorption spectroscopy. The presence of Cu^{+2} at low concentration levels (~0.01%) completely quenches the Cu^{+1} fluorescence. Several fluorohafnate compositions have been shown to be suitable hosts in regard to stability of the Cu^{+1} ion and complete elimination of Cu^{+2}. The addition of a small percentage of AlF_3 permits ease in casting large sized specimens having high optical quality. A laser rod of dimensions 5 mm diameter x 30 mm length was made having 0.3% Cu^{+1}.

The absorption and excitation spectra of Cu^{+1} were investigated to define pump requirements for a laser. Figure 2 shows the absorption spectra of Cu^{+1} in silicate and fluorohafnate glasses and NaCl crystal. The absorption spectrum of Cu^{+1} in the fluorohafnate glass showed two pump bands: the stronger band was centered at 270 nm and the weaker at 337 nm. The fluoride glasses, in general, have good UV transmission down to 230 nm. For silicate glass, however, the 270-nm absorption peak of Cu^{+1} is masked by the host absorption of the glass. The excitation spectrum of the silicate glass revealed the same two pump bands as for the fluoride glass, but with the excitation effectiveness of the 270-nm band greatly reduced as expected. For comparison, the absorption spectrum is shown for Cu:NaCl [1], demonstrating the identical absorption bands. The monovalent fluorescence in fluorohafnate glass peaked at 487 nm, a spectral shift to the blue by 50 nm from the fluorescence in silicate glass, as shown in Fig. 3. The lifetime of Cu^{+1} fluorescence was

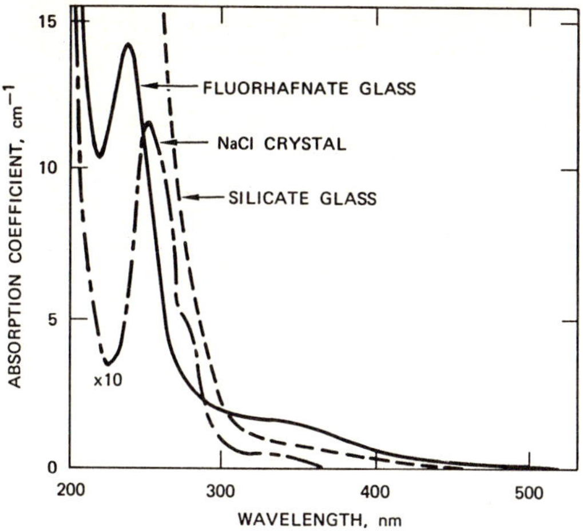

Fig.2 Absorption spectra of Cu^{+1} in silicate and fluorohafnate glass compared to Cu:NaCl

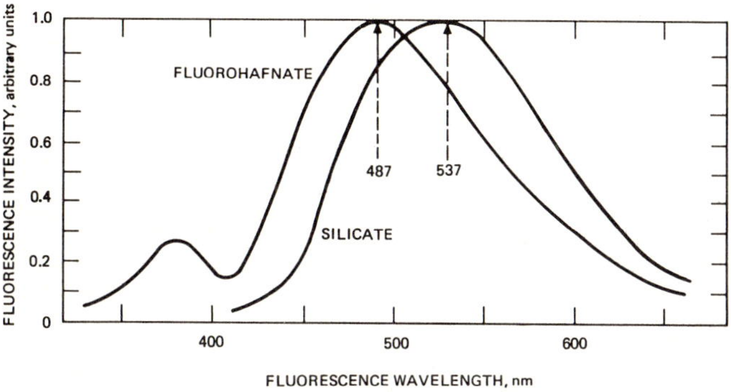

Fig. 3 Fluorescence spectra of Cu^{+1} in silicate and fluorohafnate glass

75 µs in silicate glass, and 49 µs in fluorohafnate HBT glass at room temperature.

3. Laser-pumped Laser Amplifier Experiment

At first, a laser-pumped Cu laser experiment was conducted using the quadrupled Nd:YAG (265 nm) as pump for a 30-mm long Cu fluorohafnate glass rod. No laser action was observed. We then modified the setup to a laser amplifier experiment shown in Fig. 4 to measure the gain/loss characteristics of the glass when pumped by the 265-nm beam. The probe was either a cw argon ion laser (488, 514.5 nm) or a He-Ne laser (632.8 nm), which detected the gain or loss in the sample produced by the pump. Figure 5 shows the results of this experiment, where a 2% gain was observed at 632.8 nm. However, almost no effect was observed at 514.5 nm, and a 4% loss was measured at 488 nm, which is at the peak of the fluorescence curve where the gain should be the largest. The absorption loss at 488 nm decreased to zero with a long lifetime (~3 ms) indicating that the difficulty was not excited state absorption but the formation of transient color centers by the 265-nm pump beam. This was verified by repeating the experiment with undoped fluorohafnate HBT glass, which clearly demonstrated a large transient absorption under 265-nm irradi-

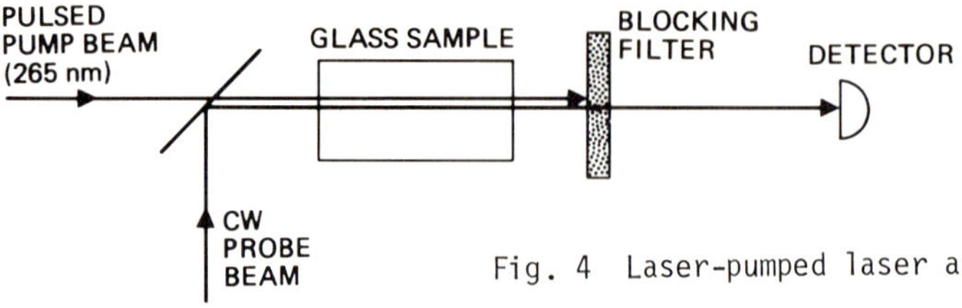

Fig. 4 Laser-pumped laser amplifier experiment

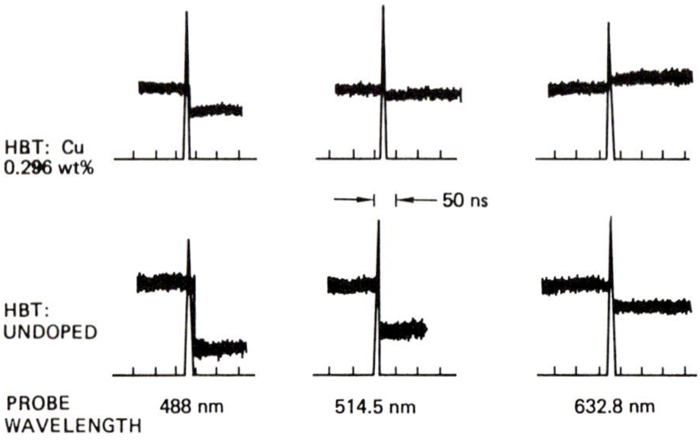

Fig. 5 Gain/loss in fluorohafnate glass laser amplifier

ation which decreased from blue to red. This transient solarization effect is consistent with the result of SIBLEY [5] where the absorption of color centers in HBL fluoride glass peaked near 330 nm and monotonically decreased with increasing wavelength. Absorption by these transient color centers undoubtedly prevented laser oscillation in the initial Cu laser experiment.

4. Monovalent Copper Ions in Crystals

Several crystals were investigated to find a host for monovalent copper that are resistant to solarization by the pump radiation. $RbMgF_3$ was investigated as a host for Cu^{+1} and proved to be nearly ideal except dopant concentrations were lower than desired (less than 0.1 atomic %). $RbMgF_3$ single crystals were grown up to 4-cm long by a seeded Czochralski procedure utilizing HF gas as a reactive atmosphere. Figure 6 shows the Cu^{+1} fluorescence for two dopant levels in $RbMgF_3$ crystals. For a dopant level ~0.01 wt % monovalent copper exhibited a strong fluorescence appeared blue to the eye, while fluorescence from another $RbMgF_3$ crystal with a five-fold increased doping of Cu^{+1} appeared green in color. This color change with increased concentration was due to the greatly increased dominance of the 575-nm fluorescence band over the 450-nm sideband. The two emission bands are attributed to Cu^{+1} ions at two inequivalent sites [6].

In addition, a potentially important oxide crystal family has been identified as a host for monovalent copper, the vanadate garnet family. One composition in this family useful for incorporation of monovalent ions is $NaCa_2Mg_2V_3O_{12}$ where Na and Ca occupy dodecahedral sites and Mg occupies the octahedral site. Differential thermal analysis showed its melting point to be about 1170°C. When 2% Cu^{+1} was doped into the Na site, a strong green fluorescence was observed peaked near 530 nm. The lifetime of this fluorescence was measured to be 3.2 μs.

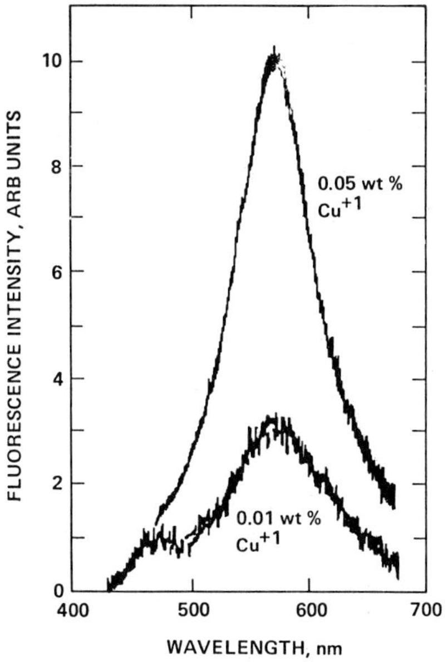

Fig. 6 Monovalent copper fluorescence in $RbMgF_3$ crystals

5. Acknowledgments

The author wishes to acknowledge the assistance of M. Robinson in producing the fluoride glasses and $RbMgF_3$ crystals, and B. A. Wechsler in making the vanadate garnet.

REFERENCES

1. A. M. MacMahon, Z. Physik 52, 336 (1928)
2. C. E. Moore, Atomic Energy Levels, vol. 2 (National Bureau of Standards, Washington, DC 1971) p. 115
3. S. Parke and R. S. Webb, Phys. Chem. Glasses 13, 157 (1972)
4. M. Robinson et al., Mat. Res. Bull. 15, 735 (1980)
5. W. A. Sibley, unpublished
6. K. Tanimura, W. A. Sibley and L. G. DeShazer, "Optical properties of Cu^+ ions in $RbMgF_3$ crystals," to be published Phys. Rev.

Part III

Vibronic Laser Materials Growth

Transition Metals in Oxide Hosts

G.M. Loiacono

Philips Laboratories, Briarcliff Manor, NY 10510, USA

Recent progress in vibronic lasers has resulted in renewed interest in bulk crystal growth techniques suitable for material preparation. The basic bulk growth processes of Bridgman, Heat Exchanger Method and Czochralski are reviewed and evaluated with respect to laser materials. The optimum operation of laser materials is dependent on a number of important crystal growth parameters, which are identified and methods of control discussed. The basic problem with incorporation of transition metals in oxide hosts is the retention of the desired oxidation state of the metal ion during the growth process. Techniques such as atmosphere control and charge compensation are reviewed.

1. INTRODUCTION

Interest in vibronic lasers has resulted in renewed efforts to identify potentially useful ion-host materials and to prepare single crystals of high optical quality for device applications [1,2]. During the early 1960's large experimental efforts existed at many laboratories and the results of these studies should be reviewed and applied to present problems.

An excellent comprehensive review of known laser materials was recently published by KAMINSKII [3]. The basic bulk growth processes for oxide materials have been reviewed by many authors and only a select few are cited in this paper [4-8]. Recent developments have included the use of automatic diameter control systems to the Czochralski method and the impressive crystals growth by Heat Exchanger Method (HEM).

The purpose of this paper is to review the available crystal growth methods, compound preparation techniques, critical parameters affecting the operation of a laser material (as related to growth) and hopefully be of use in helping to solve some present materials problems.

2. CRYSTAL GROWTH METHODS

BRIDGMAN-STOCKBARGER

This method is based on the slow solidification of a melt (contained in a suitable crucible) by either controlled lowering of the crucible through a temperature gradient or by keeping the crucible and furnace stationary and using controlled temperature lowering [9,10]. Details of this method are given in standard references [4-6]. The method has been used almost exclusively for the growth of metal halide crystals - in particular fluorides [11,12].

In evaluating crystals produced by this growth method, a number of problems are apparent, for example:

a) reaction of the crucible with the molten material,

b) multiple nucleation which hinders or prevents growth of large crystals,

c) considerable strain in the grown crystals due to crucible containment.

However, while the method is not considered suitable for the growth of large crystals, it is an excellent choice for the rapid preparation of spectroscopic study samples.

HEAT EXCHANGER METHOD (HEM)

This method is based on directional solidification of a melt by controlling the temperature gradient in the solid by a heat exchanger [13-15]. A detailed discussion of the method and its achievements is presented by Schmid at this conference.

While HEM has some similarity to the Bridgman-Stockbarger method, the differences are of significant importance and render the method almost ideally suited for large scale production of a number of materials. In the opinion of this author, HEM could be the method of choice for the large-volume, low-cost production of suitable materials.

CZOCHRALSKI METHOD

Together with the Bridgman-Stockbarger method, the Czochralski method is one of the standard growth techniques used for laser rod production. In addition to the general crystal growth references previously cited, a large number of specific reports on interesting compounds exist; for example NASSAU'S paper on $CaWO_4$:Nd [16]. While the essentials of the method have not changed over the past 67 years, major recent improvements are the development of automatic diameter control systems and control of thermal gradients by utilization of heat pipe technology [17-19]. The stimulus for these advances can be attributed to the efforts made in the perfection of semiconductor crystals (Si, GaAs). Of specific interest in oxide growth are the automated diameter control methods based on seed and crucible weighing and optical (IR, TV) methods [20-24]. A simple system presently in use for oxide growth, at Philips Laboratories, is shown in Fig. 1.

This method is especially versatile in that potentially useful compounds can be quickly surveyed for spectroscopic studies and large high optical quality crystals suitable for laser operation can be grown. In addition, the method is capable of processing production quantities of materials.

3. PREPARATION of COMPOUNDS

A standard method of compound preparation is to mix stoichiometric quantities of reactive compounds and then calcine at various temperatures until a single phase compound is formed. A typical example would be the preparation of $CaWO_4$ via the reaction.

Fig. 1. Video automatic diameter control

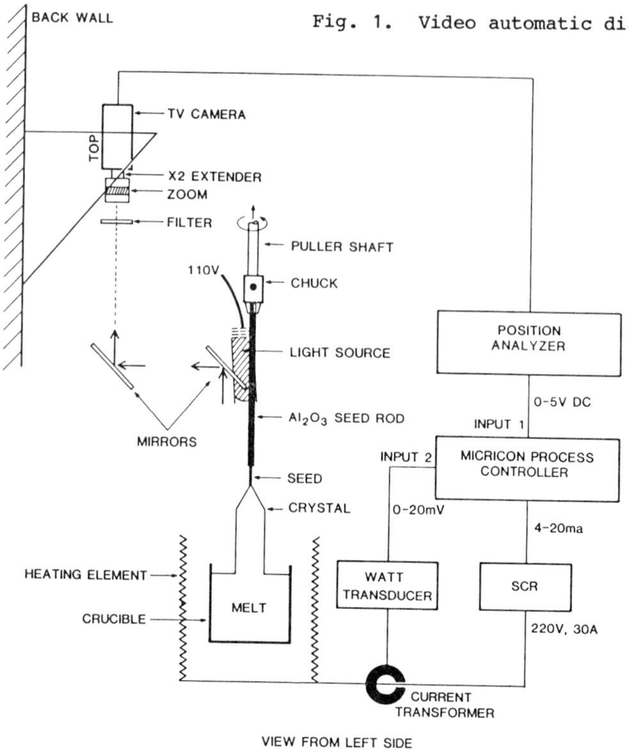

$$CaCO_3 + WO_3 \xrightarrow{\Delta} CaWO_4 + CO_2 \uparrow . \qquad (1)$$

Considerable care must be taken during processing to insure that highly volatile components (i.e., WO_3) are not lost. In addition, complications may arise due to the formation of intermediate stable compounds which inhibit formation of the primary phase.

An alternative route to the synthesis of high purity, stoichiometric compounds is via chemical precipitation methods. For example, $PbMoO_4$ was formed by aqueous solution by the reaction [25]

$$(NH_4)_2MoO_4 + PbX_2 \longrightarrow PbMoO_4 + 2NH_4X \quad , \qquad (2)$$

where $X = NO_3^- \text{ or } C_2H_3O_2^-$.

Other examples of compound formation via chemical methods can be found in the literature [26-28].

4. CRITICAL MATERIAL PARAMETERS

The successful operation of a material as a laser is dependent on a number of factors directly related to the crystal growth process. Assuming that high purity chemicals are initially used, then three inter-dependent parameters, which cause defects, must be considered: dopant distribution, crystallographic, and optical quality.

DOPANT DISTRIBUTION

The selection of an active ion/host combination is primarily determined by the size and charge of the ion in comparison to the expected substitution site in the host. Differences in size and charge misfit which can be accommodated by the lattice are on the order to ±25% and ±2, respectively [29]. For standardization and convenience in comparisons of published studies, it is suggested that the values of ionic radii and charge be taken from the work of SHANNON and PREWITT [30].

The dopant distribution in a crystal is dependent on the distribution coefficient (k), crystallographic orientation, solid/liquid interface shape, thermal gradients, etc. A detailed discussion of the equilibrium and effective distribution coefficients (k_o and k_e, respectively) has been given by LAUDISE [5]. For most cases, k_e suffices and for $1<k_e>1$, a longitudinal dopant gradient in the crystal is common. This gradient can be minimized by using large melt volumes, and only crystallizing a small fraction of the melt. In addition, radial gradients are possible and the solid/liquid interface shape must be controlled to minimize this gradient.

CRYSTALLOGRAPHIC CONSIDERATIONS

The observation of facets in the growth of some laser materials (i.e., $Y_3Al_5O_{12}$:Nd) is directly related to crystallographic orientation and thermal gradients [31]. The presence of facets during growth causes variation in dopant distribution over the facet due to differences in local growth rates. This results in the "coring" effect which leads to optical density variations in the grown crystal. The formation of facets can be eliminated by proper selection of seed orientation and maintenance of a flat solid/liquid growth interface by control of thermal gradients.

Crystallographic orientation is also important in reducing crystal misorientation (i.e., axis wander), lowering dislocation densities and minimizing optical scattering centers.

OPTICAL QUALITY

The term "optical quality" includes defects such as scattering sites, strain and refractive index variations. Scattering sites are caused by voids or inclusions where the diameter of these defects exceed 0.1 µm. High angle light scatter is caused by defect diameters of about 0.5 µm. They are formed by entrapment of gas bubbles generated from incomplete reaction of starting components or by reaction of the melt with the crucible materials. In addition, high impurity concentrations may cause the formation of complex oxides which can precipitate as solid particles in the grown crystal resulting in scattering sites and large strains.

Present processing techniques (i.e., annealing, growth using after-heaters, etc.) have eliminated strain as a problem. However, strain, resulting from large misfits during ionic substitution or very high concentrations of dopants, is not easily reduced.

Finally, low angle scattering, wave front distortion and beam divergence may be caused by refractive index inhomogeneities. Factors giving rise to index variations are crystallographic misorientations, non-uniform impurity distributions and growth-induced strains.

It is clear from this brief discussion of critical material parameters that the production of a useable laser rod, from a new material, is not a simple task. In order to be successful, a team effort is required and development time schedules are on the order of 3 to 5 years.

5. TRANSITION METALS IN OXIDE HOSTS

The incorporation of 3d transition metals (other can Cr^{+3}) in oxides has been reported [32-34]. The 4d and 5d metals have only recently been investigated and primarily in halide hosts [35].

Table 1 lists a selection of potentially useful ions recently compiled by POWELL [35]. The known ionic radii has been added, based on values from SHANNON and PREWITT for a coordination of VI [30]. A number of these ions (i.e., Cd^{+3}, Sc^{+2}, etc.) have not been assigned radii. However, as ionic charge decreases there is a corresponding increase in ionic radius and examination of available data for other ions indicates an increase of about 10 to 20%. Based on this, estimated values are given in brackets in Table 1.

Table 1. Transition metal ions †
[coordination = VI]

3d	Ionic* Radius (Å)	4d	Ionic Radius (Å)	5d	Ionic Radius (Å)
Co^{+2}	0.735	Cd^{+3}	[0.9]	Ir^{+2}	[0.8]
Co^{+3}	0.61	Pd^{+2}	0.86	Ir^{+3}	0.73
Cr^{+2}	0.82	Pd^{+3}	0.76	La^{+2}	[1.2]
Cr^{+3}	0.615	Rh^{+2}	[0.7]	Os^{+2}	[0.7]
Cu^{+2}	0.73	Rh^{+3}	0.665	Os^{+3}	[0.7]
Fe^{+2}	0.77	Ru^{+2}	[0.7]	Pt^{+2}	0.80
Fe^{+3}	0.645	Ru^{+3}	0.68	Pt^{+3}	[0.7]
Mn^{+2}	0.829	Y^{+2}	[1.0]	Re^{+2}	[0.7]
Mn^{+3}	0.65	Zr^{+3}	[0.8]	Re^{+3}	[0.7]
Ni^{+2}	0.70				
Ni^{+3}	0.60				
Sc^{+2}	[0.8]				
Ti^{+2}	0.86				
Ti^{+3}	0.67				
V^{+2}	0.79				
V^{+3}	0.64				

† R.C. Powell (1984).
* Shannon & Prewitt (1969).
[] Estimated values.

In any system concerning transition metal oxides, the desired oxidation state of the ion can be retained during growth by proper selection of atmospheric conditions (i.e., oxidizing, reducing, neutral). However, most of the ions of interest (Table 1) require reducing or neutral conditions, which then limits the choice of host material. Many potential hosts cannot be grown under these conditions (i.e., $CaMoO_4$, $LiNbO_3$) due to the reduction of one of the components. For example, $Mo^{+6} \longrightarrow Mo^{+5}$ and $Nb^{+5} \longrightarrow Nb^{+4}$. Thus a major consideration in host selection for transition metals is the question of compound stability under the required growth conditions. Oxides such as $Y_3Al_5O_{12}$, Al_2O_3 and $BeAl_2O_4$ are grown

Table 2. Potential oxide hosts for transition metals

Oxide	Crystal Structure	M.P. (°C)
$LaAlO_3$	Cubic	1940 ± 25°
Mn_2SiO_4	Orthorhombic	1345 ± 10°
$Mn_3Al_2(SiO_4)_3$	-	1195 ± 10°
$K_2MgSi_5O_{12}$	Cubic	1090 ± 10°
$Cs_2BeSi_5O_{12}$	Cubic	1430 ± 10°
$CaMgSi_2O_6$	Monoclinic	1390 ± 10°

under argon or nitrogen atmospheres and therefore are excellent choices. A few additional oxide candidates (Table 2) have also been examined and found satisfactory. All of the compounds listed are congruently melting.

Another method of stabilizing oxidation states is by coupled substitution which results in a number of charge transfer mechanisms. Discussions of these effects and mechanisms are given in many mineralogy texts [36,37]. For example, in Al_2O_3, Ti^{+3} can be stabilized by coupled substitution of Fe^{+2} and Ti^{+4} which results in intervalence charge transfer of the form

$$Fe^{+2} + Ti^{+4} \rightleftharpoons Fe^{+3} + Ti^{+3} \qquad (3)$$

Other examples are the stabilization of Co^{+2} in Al_2O_3 by Ti^{+4} according to [38]

$$Co^{+3} + Ti^{+3} \rightleftharpoons Co^{+2} + Ti^{+4} \qquad (4)$$

and of Ru^{+2} in TiO_2 by Ta^{+5} [39].

The results of our preliminary experiments with the host crystals $KTiPO_5$, $Bi_4Ge_3O_{12}$, and Mn_2SiO_4 were recently summarized [40]. The most promising material was Mn_2SiO_4 which exhibited strong luminescence at 300°K near 700 nm with a lifetime of 3.6 μsec. Recent growth efforts have concentrated on dilution of the [Mn] with Mg and Ca and measurements by R.C. Powell are presently in progress.

6. CONCLUSIONS

This brief summary on the crystal growth, material parameters and oxidation state stabilization of potential solid state lasers, indicates a need for well coordinated research efforts. Close cooperation and communication between workers in the critical scientific disciplines are essential in order to discover and develop new materials for laser applications. It should be noted that many published studies have been reported over the past 25 years, of which the author has only referenced a fraction.

1. P.T. Kenyon, L. Andrews, B. McCollum and A. Lempicki, I.J.Q.E. QE-18 (8), 1189 (1982).
2. J.C. Walling, O.G. Peterson, H.P. Jennsen, R.C. Morris, and E.W. O'Dell, I.J.Q.E. QE-16 (12), 1302 (1980).

3. A.A. Kaminskii, Laser Crystals, Springer-Verlag, Berlin (1981).
4. J.C. Brice, The Growth of Crystals from Liquids, North Holland Publ. Co., Amsterdam (1973).
5. R.A. Laudise, The Growth of Single Crystals, Prentice-Hall, N.J. (1970).
6. Crystal Growth and Characterization, Ed. by R. Ueda and J.B. Mullin, North Holland Publ. Co., Amsterdam (1975).
7. Crystal Growth, Ed. by B.R. Pamplin (2nd Ed.) Pergamon Press, N.Y. (1980).
8. K. Th. Wilke, Kristallzuchtung, VEB Deutscher, Verlag Wissen., Berlin (1973).
9. P.W. Bridgman, Proc. Amer. Acad. Sci. 60, 305 (1925).
10. D.C. Stockbarger, Rev. Sci. Instr. 7, 133 (1936).
11. H. Guggenheim, J. Appl. Phys. 34, 2842 (1963).
12. W.D. Scott, J. Am. Ceram. Soc. 45, 586 (1962).
13. D. Viechnicki and F. Schmid, J. Crystal Growth 11, 345 (1971).
14. D. Viechnicki, F Schmid and J.W. McCauley, J. Appl. Phys. 43, 4805 (1972).
15. J.L. Caslavsky and D. Viechnicki, J. Crystal Growth 46, 601 (1979).
16. K. Nassau and A.M. Broyer, J. Appl. Phys. 33, 3064 (1962).
17. T. Swiek, B. Chalmers, and A.I. Malvski, J. Crystal Growth 42, 453 (1977).
18. W. Bardsley, D.T.J. Hurle, G.C. Joyce and G.C. Wilson, J. Crystal Growth 40, 21 (1982).
19. Z.J. Xing, J.K. Kenney and A.F. Witt, J. Crystal Growth 59, 659 (1982).
20. T.R. Kyle and G. Zydzik, Matl. Res. Bull. 8, 443 (1973).
21. K.J. Gartner, K.F. Rittinghaus, A. Seeger and W. Uelhoff, J. Crystal Growth 13/14, 619 (1972).
22. D.F. O'Kane, T.W. Kwap, L. Gulity and A.L. Bednowitz, J. Crystal Growth 13/14, 624 (1972).
23. U. Gross and R. Kersten, J. Crystal Growth 15, 85 (1982).
24. A.E. Zinnes, B.E. Nevis and C.D. Brandle, J. Crystal Growth 19, 187 (1973).
25. G.M. Loiacono, J.F. Balascio, R. Bonner and A. Savage, J. Crystal Growth 21, 1 (1974).
26. L.H. Brixner, J. Electrochem. Soc. 111, 690 (1964).
27. D.R. Messier and G.E. Gazza, Ceram. Bull. 51, 692 (1972).
28. E. Camall and D. Pearlman, Matl. Res. Bull. 7, 647 (1972).
29. K. Nassau and G.M. Loiacono, J. Phys. Chem. Solids 24, 1503 (1963).
30. R.D. Shannon and C.T. Prewitt, Acta Cryst. B25, 925 (1969).
31. C.D. Brandle and A.J. Valentino, J. Crystal Growth 12, 3 (1972).
32. P.A. Arsenev and B.A. Baranov, Phys. Stat. Sol.(a) 9, 673 (1972).
33. V.A. Antonov and P.A. Arsenev, Phys. Stat. Sol.(a) 20, K157 (1973).
34. V.A. Antonov, P.A. Arsenev, S.A. Vakhidov and E.M. Ibraginova, Phys. Stat. Sol.(a) 22, 391 (1974).
35. R.C. Powell, private communication (1984).
36. G. Smith and R.G.J. Sterns, "The Physics and Chemistry of Minerals and Rocks", Ed. R.G.J. Sterns, John Wiley & Sons, N.Y. 583 (1976).
37. W. Eitel, "The Physical Chemistry of Silicates", Univ. Chicago Press, Illinois 69 (1954).
38. G.A. Keig, J. Crystal Growth 2, 356 (1968).
39. P. Triggs, F. Levy and F.E. Wagner, Mat. Res. Bull. 19, 197 (1984).
40. S. Colak, G.M. Loiacono, G. Mizell, W. Zwicker and R.C. Powell, Proc. Intern. Conf. Lasers 1983, in press (1984).

Growth of Crystals for Solid State Lasers

Milan R. Kokta

Union Carbide Corporation, 750 South 32nd Street, Washougal, WA 98671, USA

1. Abstract

The crystal growth of high melting point oxide crystals by the Czochralski technique is described. The group of crystals was selected with emphasis on laser hosts and includes Ti^{3+}-doped sapphire, gadolinium scandium gallium garnet, and its aluminum analog, yttrium aluminum garnet. Material preparation, growth experiments design considerations, consideration for scaling up the crystal size, and crystal evaluation are included. Results of work leading to growth of 3" diameter YAG crystals are presented. The paper presents a general approach to crystal growth including crucible material selection, atmosphere control, and fluid dynamics considerations.

2. Introduction

The need for producing high optical quality, defect-free crystalline materials for use as optical and electronic components led to the development of many crystal growing techniques.

The basic methods were developed, modified, and refined to enhance their applicability to specific materials or groups of materials, as well as the desired quality of bulk crystalline solid. Solid state electronics and optics are widely based on single crystal materials and their progress forced a variety of studies aimed at understanding and improving the final product – the crystal.

Rich sources of information now exist in specific areas of thermodynamics of liquids, fluid dynamics, nucleation, growth rates, segregation, interface morphology, and stability and crystalline defects. However, because of the complex nature of the processes involved in crystal growth and the wide variety of variables controlling these processes, a large gap between the theories and practical application of crystal growth persists.

The growth of any particular material requires careful selection of the proper technique based on the material properties and the required quality. In addition, the simple matter of scaling the crystal size may result in a combining of various disciplines of science to reach the goal. Categorizing the crystal growth leads up to three basic categories:

1) solution growth
2) melt growth
3) vapor phase growth.

Solution growth consists of

1) aqueous solution growth
2) molten salt growth (flux)
3) metallic solution growth.

Solution growth methods generally do not provide laser quality material in bulk form.

The melt growth techniques are applicable for materials which melt and solidify congruently. Such techniques include

1) Bridgman techniques
2) zone melting
3) edge defined techniques
4) skull melting
5) crystal pulling — Czochralski
6) heat exchange method.

In the growth of laser quality oxide crystal, the Czochralski technique is still prevalent and produces most of the commercial solid state laser material.

3. Crystal Pulling

The crystal pulling technique is one of the most versatile and widely used of the variety of methods in use today. The method was introduced by CZOCHRALSKI [1] and is now the dominant process used in industry today for the production of semiconductor and oxide crystals.

The process is simple in principle. Material contained in a suitable crucible is heated until the charge is melted. The thermal environment is adjusted so the equilibrium between liquid and solid is approached on the surface of the melt in the crucible center. A seed crystal is placed into the "cold spot" and, by controlling the melt temperature, solidification may be started. If the seed is rotated and withdrawn (pulled) from the melt surface at a rate not exceeding the rate of solidification, a cylindrical crystal known as a "boule" or "ingot" is the result. The crystal diameter control is usually achieved by controlling the liquid temperature.

Although simple in principle, the pulling method is complex in detail and a wide variety of variables must be carefully controlled. The consideration of applicability starts with evaluation of the nature of material to be grown, because the method does not lend itself easily to growing materials 1) incongruently melting, 2) exhibiting high vapor pressure at the melt temperature, or 3) materials with a first order solid-solid phase transition.

The following is the consideration of crucible materials and the type of heat source. The crucible has to be selected on the basis of compatibility with the melt; melting point of the crucible, mechanical properties, chemical stability, and type of heating determine the selection. The size of the crucible is dictated by the desired crystal size, and heat and fluid flow geometries [2].

The most commonly utilized heat sources are resistance heaters and induction heating. Resistance heating is normally limited to low temperature applications. In Table 1 below, some of the materials used in resistance heating are compiled [3].

For growth of high melting point compounds, especially oxides, induction heating is commonly utilized. Inductive or radiofrequency heating utilizes the coupling of oscillating power through the induction coil into the susceptor. The susceptor has to be conductive and it is heated to the required temperature. Radiofrequency heating operates in a frequency range from 200 to 500 kHz and a power range of 20 to 100 kw.

Table 1. Resistance elements

Material	Useful temperature [°C]	Atmosphere
Nichrome	⩽ 1200	Air
SiC	⩽ 1500	Air
Pt-50%	⩽ 1800	Air
Carbon	⩽ 3000	Neutral, reducing
Molybdenum	⩽ 2400	Vacuum, neutral, reducing
Tungsten	⩽ 3000	Vacuum, neutral, reducing

Very often the crucible serves directly as susceptor and then becomes the heater. This arrangement offers a direct way of controlling the thermal conditions of the molten charge, by changing the geometrical position of the crucible within the induction coil. Table 2 from reference 3 compiles the most commonly utilized crucible materials.

The growth furnace construction allows for changes in thermal configuration of the system, that is thermal gradients. These can be internal —within the liquid bulk, or external —above the crucible (vertical and horizontal). Internal thermal gradients are influenced by crucible position, crucible shape, and furnace construction (quality and quantity of insulation used). External thermal gradients are mainly influenced by furnace construction (using passive or active afterheaters, baffles, etc., see Fig.1). Examples of furnace constructions are shown by BRANDLE [3].

Table 2. Common Crucibles

Material	Maximum Operating Temperature [°C]	Melting Point [°C]	Use
Platinum	1400	1773	1,2
Iridium	2150	2452	1,2
Molybdenum	2300	2620	1,2
Tungsten	2800	3370	1,2
Carbon	3000	-	1,2
Silica	1550	1700	1
Alumina	1800	2050	1
1 = resistance		2 = r.f. heating	

4. Crystal Growth Parameters

The success in pulling crystals using the Czochralski technique depends on the right combination of numerous variables. The pulling rate, rotation rate, thermal geometry, mass flow, heat flow, and melt size have effects on crystal perfection.

The pulling rate determines the growth rate of the crystal and is the most independent variable. The growth rate is dictated by material properties, mainly by the kinetics of its solidification. The quality of the re-

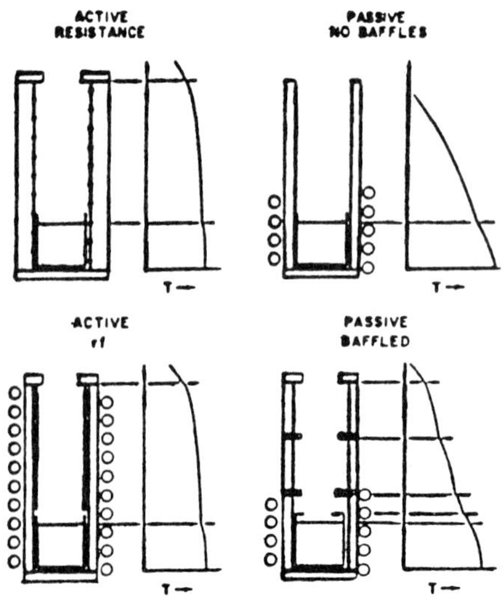

Fig. 2

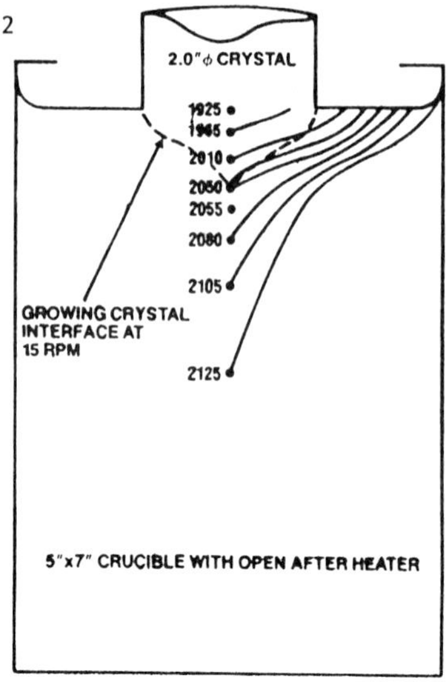

Fig.1. Different types of afterheaters and thermal gradients produced

Fig.2. Interface shape and shape isotherms

sulting crystals is strongly influenced by growth rates with regard to inclusion, bubbles, and internal strain. The growth rate of crystals in a Czochralski set-up is defined as the effective linear growth rate which is the rate of the advance of the solid-liquid interface. The linear growth rate may be defined by

$$G = \left(1 + \frac{\rho_s r^2}{R^2 \rho_1 - r^2 \rho_s}\right),$$

where G is the linear growth rate, P is the pull rate, r is the crystal diameter, R the crucible diameter, and ρ_s, ρ_1 are the liquid and solid densities.

The remaining variables such as rotation rates, atmosphere, thermal gradients are combined in order to arrive at the desired interface shape. The interface shape adopts the shape isotherm existing in the bulk of the melt (see Fig.2).

The isotherm depends on convection currents existing in the fluid from which the crystal is being withdrawn. There are two major types of convection:

1) Natural convection, which is driven by buoyancy forces which exist due to density changes through the liquid created by temperature gradients, and gravity.

2) Forced convection, which is driven by forces caused by rotating the crystal in the fluid.

The convection and fluid flow in Czochralski growth systems was treated in numerous works [4-17].

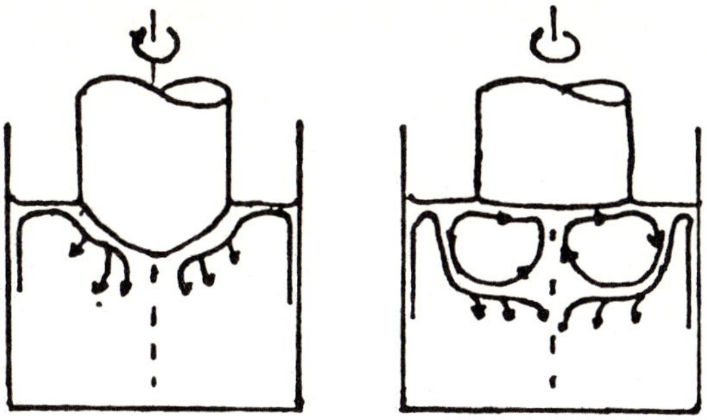

Fig.3. Shapes of solid-liquid interfaces

For practical purposes, Fig.3 indicates the interface shape resulting from the fluid flow dominant under various conditions.

The determination of the fluid convection dominant under given conditions is illustrated schematically by BRANDLE [3] and is shown in Fig.4. The flow patterns are presented as they may be viewed on the surface of the melt. The Type II flow produces a flat interface which is desirable because the facet formation on the bulk of the crystal intersection is limited. Facets are moved towards the surface of the crystal and thus, better crystal utilization may be achieved.

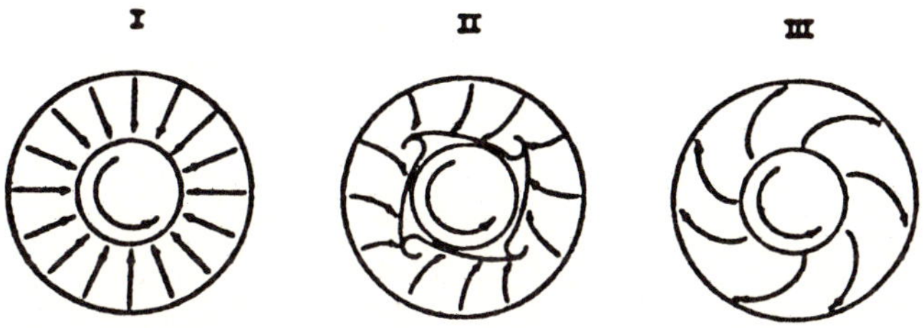

Fig.4. Simulated fluid flow in crystal pulling system; (**a**) slow rotation; (**b**) moderate rotation; and (**c**) fast rotation

The preferable mode of growth is growth with a flat interface (Type II) in order to avoid facet strain. Faceting and resulting strain is observable in most oxide crystals and is best documented for <111> direction growth in garnet crystals (Fig.5).

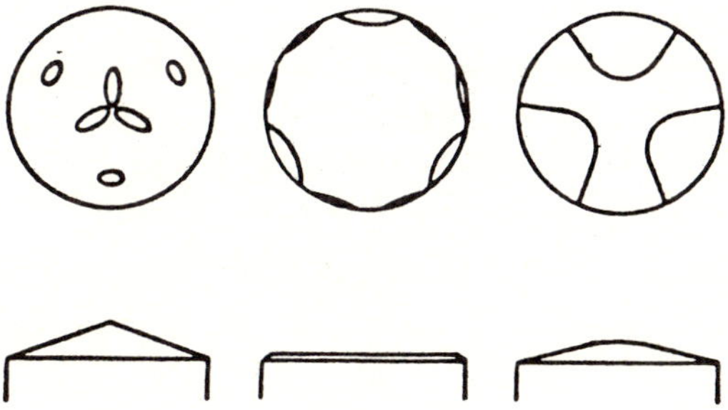

Fig.5. Schematic showing facets formed on growth interface of a <111> garnet crystal

Changing the facet position can be accomplished by changing the interface shape, by changing temperature gradients, or a balance between natural and forced convective flow. CARRUTHERS [4] showed that inversion of the interface is dependent on the crystal diameter d, the melt expansion coefficient $\propto$, the radial thermal gradient ΔT, the crucible radius R, and the crystal rotation ω. These variables provide a critical inversion diameter given by

$$d_c = [g\propto\Delta T R^3 \pi^{-2}]^{\frac{1}{4}} \omega^{-\frac{1}{2}} \quad ,$$

where g is the gravitational acceleration.

However, it is not always practical to attempt the flat interface growth, especially in cases where maintaining the flat interface stability is extremely difficult due to the fact that the melt composition or the thermal environment changes during the course of the growth process. Such problems are often encountered in the growth of crystals doped with active ions which have limited solubility in the host material, such as Nd-doped YAG or Ti^{3+}-doped sapphire. The concentration of the dopant ion is given by $C_S = kC_0(1-g)^{k-1}$ where C_0 is the initial concentration of the dopant in the melt, g is the fraction of melt which has been solidified, and k is the distribution coefficient or effective distribution coefficient defined as the ratio of the concentration in the solid to the concentration in the melt at a given instant, that is

$$k = \left(\frac{C\ \text{solid}}{C\ \text{melt}}\right)_t \quad .$$

The distribution coefficient may be used to calculate the concentration of dopant in the crystal, shifts in the lattice parameter, and changes in the melt stoichiometry during the crystal growth run. Because of compositional changes, it is advisable to maintain growth conditions leading to a convex interface for a crystal with low distribution coefficients such as Nd:YAG ($k = 0.18$), where a more stable interface can be maintained. In such cases, in order to decrease facet strain and dopant segregation effects, it is advisable to grow crystal boules of large diameter. An example, results of passive interferometry on a Nd:YAG rod taken from a 3" diameter boule showing virtually no wavefront distortion, is presented in Fig.6.

5. Scaling the Crystal Sizes

New technologies in optical applications are just beginning to develop, some of which may and will require scale-up of physical dimensions of crystalline materials. As an example, the development of slab laser configuration plates

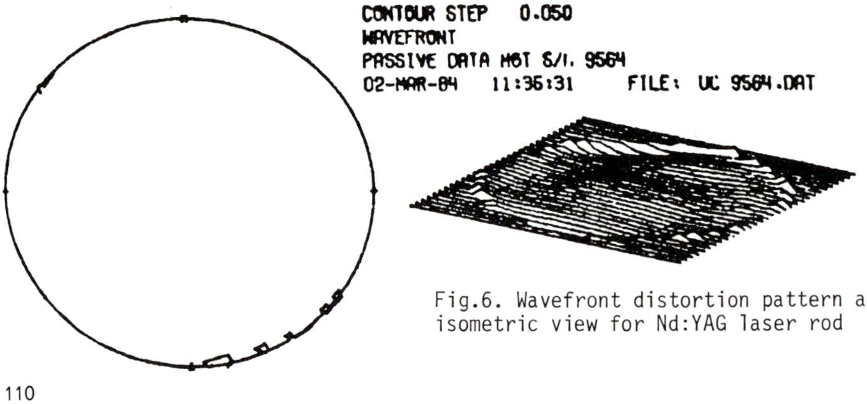

Fig.6. Wavefront distortion pattern and isometric view for Nd:YAG laser rod

increased demands on crystal size. The most important factors influencing the results are 1) pull and rotation rates, 2) thermal gradients, and 3) material considerations. A detailed review of scale-up problems is presented by BRANDLE [18]. The conditions for growth of large diameter boules may be calculated; a summary of such data calculated for large diameter YAG boules is given in Table 3.

Table 3. Conditions for Growth of Large Diameter YAG Boules

5" × 7" Crucible:			
Rotation Rate	8	10	15
Re	430	537	806
Di ["]		2.55	2.08
6" × 8" Crucible:			
Rotation Rate	8	10	15
Re	967	1210	1815
Di ["]	3.12	2.8	2.28

Re = Reynolds number
Di = Inversion diameter

From the above abbreviated description, it is apparent that the pulling technique is a powerful tool used in the manufacture of single crystals. Its basic weakness, which is complexity and number of variables, is also its biggest strength, providing the crystal grower with enormous versatility and numerous approaches to a given task.

One of the most elegant recent variations is a technique for pulling crystals from a directly heated melt, reported recently by D. MATEIKA et al. [19]. This opens a wide variety of applications previously impossible because of the lack of suitable crucibles.

6. Practical Examples

a) Growth of Garnet Crystals

Crystals grouped under the common name "garnets" offer large flexibility for doping because the three non-equivalent cation sublattices allow for various substitutions and lattice parameter changes. Recently, garnets have enjoyed renewed interest [20-22]. There are basically three groups of garnet compounds which may be successfully grown by pulling techniques:

1) Gallium oxide based garnet, mainly derived from gadolinium gallium garnet (GGG)
2) Aluminum based garnets, derived from yttrium aluminum garnet (YAG)
3) Germanate garnets, a group which so far has received least attention because of the difficulties encountered in growth related to the high vapor pressure of GeO_2.

The gallium based garnet offers flexibility in octahedral site substitutions. Growth of it is similar to GGG growth, is relatively uncomplicated, and a large data base exists for growth of GGG. GGG growth can yield large (4" diameter) boules with flat interfaces. The disadvantage of these materials is that they easily incorporate impurities related to iridium and its charge compensation. The presence of oxygen vacancies was also documented by STACY et al. [23] and STACY [24]. The existence of color centers in GGG has not yet been satisfactorily explained.

Aluminum based YAG is being grown for laser applications in large quantities. Its derivative, the scandium doped $Gd_3Sc_2Al_3O_{12}$, was first grown in 1973 in Bell Labs by BRANDLE [20]. Recently, this material was grown by Union Carbide in codoped Nd-Cr form [26], and singly doped with chromium. A disadvantage of aluminum based garnets with respect to growth difficulty is maintaining stable flat interface growth. Easy formation of fluid flow instabilities is caused by relatively high destabilizing vertical gradients in the flow region, requiring balance between natural and forced convection. Changes in fluid flow pattern lead to crystal melt-back on regrowth resulting in the generation of dislocation, inclusion, and voids entrapment.

Germanium garnets offer the advantage of a large range of lattice parameters, and thus variation of the crystal field strength. The major disadvantage is the high volatility of germanium oxides, especially GeO, which is formed by dissociation in oxygen-deficient ambient atmospheres. The ambient atmosphere is usually maintained inert with a small O_2 concentration only in order to protect crucible materials such as iridium.

The new pulling technique from the directly heated melt, reported recently by MATEIKA et al. [19], seems to be tailor-made for the growth of germanate garnets. Germanate garnets worthy of investigation are summarized in Table 4.

The onset of laser damage of gadolinium scandium gallium garnet grown by the Union Carbide Corporation was recently measured by Del Dobberpuhl and S. Feitzel of the Naval Weapons Center in China Lake, and was compared with YAG (Table 5).

b) Growth of Other Oxide Crystals

Other oxide crystals for laser application, such as alexandrite, are also grown by the pulling technique. One of the old crystals, corundum, grown as ruby for lasers, is finding new applications for tunable lasers in the 730 to 840 nm range.

Table 4. Germanate garnets suitable as laser host

Octahedral ion	a_0(Å)
$\{Ca_3\}[X_2]\ (Ge_3)O_{12}$	-
Al_2	12.120
Sc_2	12.504
Ga_2	12.251
In_2	12.620
Y_2	12.805
Lu_2	12.730
$\{Sr_3\}[X_2]\ (Ge_3)O_{12}$	
Sc_2	12.785
In_2	12.870
Y_2	13.085
Lu_2	13.010

Table 5. Onset of laser damage

Material	Damage onset [GW/cm^2]	Defect density [cm^{-2}]
$Gd_3Sc_2Ga_3O_{12}$	1.08 ± 0.06	3.9×10^4
YAG standard	1.42 ± 0.32	1.10×10^4
YAG UCC	1.21 ± 0.5	1.5×10^4
$Gd_3Sc_2Al_3O_{12}$:Cr*	0.316 ± 0.132	4.7×10^3

*2% Cr^{3+} on Sc sites

The growth of Ti^{3+}:Al_2O_3 can be carried out by various methods (HEM). Results obtained on material grown by the pulling technique at Union Carbide were presented by DESHAZER [27]. It was found that the effective distribution coefficient for Ti^{3+} is extremely low (k = 0.14) for applied growth conditions. The growth rates are controlled by the dissociation kinetics of the reaction

$$2TiO_2 \rightleftarrows Ti_2O_3 + \frac{1}{2} O_2$$

which takes place at the melt surface. Therefore, control of the growth atmosphere is an important factor in controlling the amount of Ti^{3+} being incorporated into corundum structure.

7. Conclusions

It is evident that the crystal field of the laser host may be utilized to alter optical properties of active ions in solid state laser systems as may be illustrated by the case of Cr^{3+} (Table 6).

Table 6. Absorption wavelength of Cr^{3+} in various hosts

Host	Cell size [Å]	Absorption λ [nm]
$MgAl_2O_4$ (spinel)	8.086	391
Al_2O_3 (corundum)	a_o = 4.763, c_o = 13.003	402
$Y_3Al_5O_{12}$ (YAG)	12.010	432
$Gd_3Sc_2Al_3O_{12}$ (garnet)	12.420	459
$Gd_3Sc_2Ga_3O_{12}$ (garnet)	12.696	458

Therefore, a systematic investigation of potential laser hosts and techniques for their growth will yield desired laser systems.

1. J. Czochralski: Z. Phys. Chem. **92**, 219 (1918)
2. J.R. Carruthers: J. Crystal Growth **32**, 13 (1976)
3. C.D. Brandle: "Crystal Growth", Second Edition, ed. by B.R. Pamplin, Pergamon Press (1980)
4. J.R. Carruthers: J. Cryst. Growth **36**, 212 (1976)

5. K. Tagaki, T. Fukasawa, M. Ishii: J. Crystal Growth **32**, 89 (1976)
6. B. Cockayne, B. Lent, J.M. Roslington: J. Materials Science **11**, 259 (1976)
7. K. Shiroki: J. Crystal Growth **40**, 129 (1977)
8. D.C. Miller, A.J. Valentino, L.C. Shick: J. Crystal Growth **44**, 121 (1978)
9. N. Kobayashi: J. Crystal Growth **43**, 357 (1978)
10. M. Miyazawa, Y. Mori, S. Homma, K. Kitamura: Mat. Res. Bull. **13**, 675 (1978)
11. W.E. Langolis: J. Crystal Growth **46**, 743 (1979)
12. N. Kobayashi, T. Arizumi: J. Crystal Growth **49**, 419 (1980)
13. N. Kobayashi: J. Crystal Growth **52**, 425 (1981)
14. M. Miyazawa: J. Crystal Growth **53**, 636 (1981)
15. D.C. Miller, T.L. Pernell: J. Crystal Growth **53**, 523 (1981)
16. N. Kobayashi: J. Crystal Growth **55**, 339 (1981)
17. D.T.J. Hurle: J. Crystal Growth **65**, 124 (1983)
18. C.D. Brandle: Crystal Growth, a Tutorial Approach 2, ed. by W. Bardsley, D.T.J. Hurle, and J.B. Mullin (1979)
19. D. Mateika, R. Laurieu, M. Liehr: J. Crystal Growth **65**, 237 (1983)
20. C.D. Brandle, J.C. Vanderleeden: IEEE Quantum Electron., Vol.QE-10, **67** (1974)
21. E.V. Zharikov, N.N. Iichev, V.V. Laptev, A.A. Malyutin, V.G. Ostrumov, P.P. Pashinin, I.A. Shcherbakov: Sov. J. Quantum Electron.**12**, 338 (3) (1983)
22. E.V. Zharikov, V.V. Laptev, E.I. Sidorova, Yu. P. Timofeev, I.A. Shcherbakov: Sov. J. Quantum Electron.**12**, 1124 (8) (1982)
23. W.T. Stacy, R. Metslaar, P.K. Larson, A. Briell, J.A. Pistorius: Appl. Phys. Lett. **24**, 254 (1974)
24. W.T. Stacy: J. Crystal Growth **24/25**, 137 (1974)
25. D. Pruss, G. Huber, A. Beimowski, V.V. Laptev, I.A. Shcherbakov, Y.V. Zharikov: Appl. Phys. B**28**, 355 (1982)
26. L.G. DeShazer, S.C. Rand, M.L. Minden, J.H.W. Liaw, M.R. Kokta: Site abstract in J. Opt. Soc. Amer., to be published in J. Opt. Soc. Amer.
27. L.G. DeShazer: Site abstract in J. Opt. Soc. Amer., to be published in J. Opt. Soc. Amer.

Growth of Fluoride Laser Host Crystals

A. Linz and D.R. Gabbe

Massachusetts Institute of Technology, Department of Electrical Engineering and Computer Science, Center for Materials Science and Engineering, Crystal Physics and Optical Electronics Laboratory
Cambridge, MA 02139, USA

Introduction Fluoride crystals have some general characteristics that are desirable for a laser host. These include a wide transparency range, low refractive index and nonlinear index. They can in general be grown with a better optical quality than most oxide crystals. $LiYF_4$ is an example of a very successful fluoride host crystal and, even though it does not have quite as good mechanical and thermal properties as YAG, in many applications it performs as well or better than YAG. The principal conditions that must be met if a compound is to be useful are that it is the primary phase crystalizing from the melt, that the yield be practicably large, that the crystal does not undergo a phase-transformation upon being cooled to room temperature and that the melt be sufficiently stable (with regard to evaporation and oxidation state of dopants) over the several days normally required for crystal growth.

Growth of high optical quality fluoride laser host crystals is highly dependent on the use of high purity oxygen-free starting materials and rigorous attention to keeping treatment and growth furnaces free of oxygen and water vapor contamination. This paper will outline the various purification and hydrofluorination processes used in this laboratory, as well as the application of the various growth techniques.

1. Feed Preparation. To be used for growth of optical crystals the fluorides must be highly purified with regard to both anions and cations. Depending on the type of starting chemicals available and upon the goals of the crystal growth experiment, differing lines of feed purification and synthesis are followed. In the usual purification procedure a water-soluble compound of the desired metal (cation) is purified of other cations using wet chemical methods. The purified cation is then precipitated from the solution, usually as a carbonate which is then converted to the fluoride by reaction with HF gas (hydrofluorination) at temperatures up to 1450°C. During the wet chemical processing, various anion impurities such as sulfate or silicate are removed as well. However, the principal means of anion purification is hydrofluorination.

Often no wet chemical prepurification is required. In such cases a commercial chemical, (reagent grade oxide or carbonate or a fluoride obtained from a specialty chemical supplier) is hydrofluorinated directly, omitting any kind of pretreatment. In certain cases after synthesis and purification by hydrofluorination, the resulting fluoride is further purified by zone-refining. This process has found special application in the purification of rare earth fluorides in that we have been able to effect purification with respect to other rare earths.

2. Wet Chemical Purification. The goal of wet chemical purification is to process in such a way a soluble salt of the metal (cation) whose fluoride we require, so as to remove impurities offensive to crystal growth and optical properties.

To enable one to carry out the purification processes effectively the laboratory has a class 100 chemical clean room, as well as numerous pieces of equipment relating to ultra-pure chemical preparation that have been specially designed and constructed to facilitate ultra purification of kilogram quantities of materials.

The broad spectrum purification techniques of solvent extraction, ion exchange, coprecipitation, submicrometer filtration, precipitation from homogeneous solution and other specialized precipitation methods are routinely applied. If it is known that a specific impurity must be removed, then more specific chemical procedures are used.

Synthesis of purified materials is a dynamic process in which the rate of purification must be maintained higher than the rate of recontamination. This requirement necessitates the use of the above-mentioned clean room and auxillary equipment specially developed in this laboratory. In all instances, after purification of the solution, the purified metal ion is separated by formation of a precipitate which is then filtered from the supernatant liquid, washed and stored for subsequent conversion to fluoride by reaction with HF gas. Most often this precipitate is a carbonate. Methods employing formation under CO_2 pressure of complex bicarbonate solutions from which a crystalline carbonate can be recovered have been developed. Also developed has been the technique of precipitation from homogeneous solution under CO_2 pressure in which decomposition of urea leads to precipation of a well crystallized carbonate.

3. <u>Hydrofluorination</u>. Conversion of the purified carbonate, commercial carbonate or oxide to rigorously oxygen-free fluoride is done by reaction with HF gas.

For this purpose a Pt lined process tube containing a Pt boat supported by a graphite hemicylinder and fitted with Teflon end caps is heated in a three-zone furnace to a maximum temperature of 1450°C as shown in Fig. 1. HF gas drawn from a commercial cylinder is passed over the heated charge converting it to fluoride and releasing water and CO_2 which passes out of the tube. Ultimately the fluoride formed in this reaction, is melted in HF to allow complete removal of oxide. In many cases fluorides are available commercially, but are not sufficiently oxide-free to be used for crystal growth. Hydrofluorination of these materials is used to effect purification. Rigorous hydrofluorination of the molten fluorides in addition to purifying with respect to oxygen also results in removing other anions that may be present.

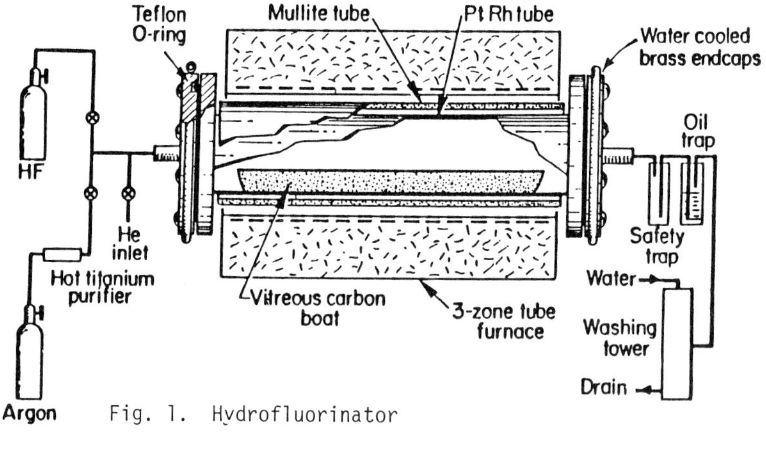

Fig. 1. Hydrofluorinator

4. Crystal Growth. There are two main growth techniques that have been found practical for the production of fluoride crystals. One is a Bridgman type process where a thermal gradient is moved through the melt such that normal freezing occurs from the bottom to the top of a crucible at an appropriate rate. One advantage is that the crucible may be sealed, thus minimizing evaporation problems. Another is that it is a basically simpler process and the equipment is less costly. However, it is difficult to obtain high-quality crystals of other than cubic materials.

The other technique is the Czochralski pulling process, which is the major one used for production of laser crystals. Once the growth parameters for a particular material have been defined, very high quality crystals may be obtained, provided the vapor pressure above the melt is not too high. It is a much more complex process than Bridgman growth, but also can yield more phase information, although at the cost of a lot more time and effort.

5. Bridgman Growth. We currently use a simple Bridgman furnace. The crucible is impervious graphite of about 1.2 cm ID, heated by an RF heated graphite susceptor. The crucible is evacuated and backfilled with Ar through a weep hole which is then sealed by a plug of solidified melt. The crucible is lowered out of the hot zone at about 1 mm/hr. A typical crystal from a series of experiments on elpasolite growth is shown in Fig. 2. While this furnace has proven to be very valuable for exploratory work on new systems, the existence of only one hot zone gives very little control over the temperature gradient and, consequently, the shape of the liquid-solid interface. For some systems this leads to constitutional supercooling and the crystals are not always suitable for accurate optical characterization. Development work on a three-zone fluoride Bridgman furnace is in progress.

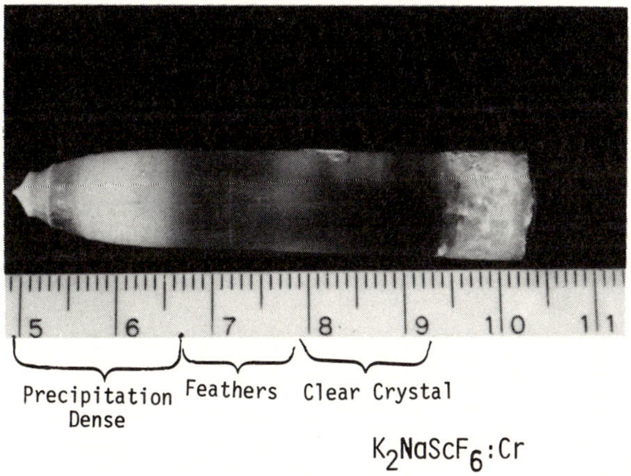

Fig. 2. Bridgman Crystal

6. Zone-refining. Zone-refining is used in purification of fluorides in those cases where other methods are not easily applied. It may also be considered a form of horizontal Bridgman growth and is useful as a relatively quick method of obtaining small crystals of congruently melting materials for fluorescent evaluation.

The zone-refiner employs a thin walled tubular graphite susceptor and a flat concentrator type RF coil to heat a narrow zone as shown in Fig. 3. A

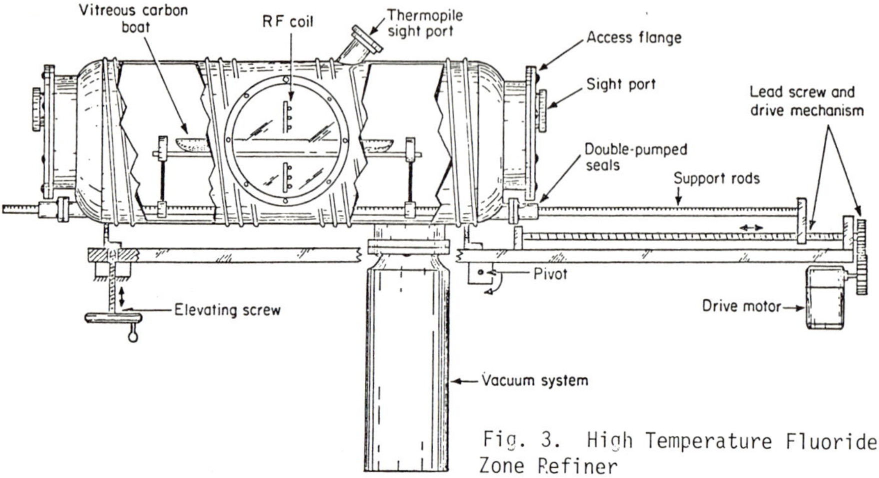

Fig. 3. High Temperature Fluoride Zone Refiner

vitreous carbon or platinum boat lies within the susceptor. This assembly is supported at the ends and is attached to a pair of movable rods sealed through the walls of the furnace chamber. Zone-refining is achieved by moving the charge through the fixed work coil. The principal successes of zone-refining have been in purification of rare earth and other fluorides, synthesis of multicomponent phases and in investigation of phase equilibria. Purification with respect to other rare earths of ErF_3, $LiErF_4$, and TbF_3 and YF_3 has been carried out as shown by fluorescence, mass spectroscopy and fluorescent lifetime measurements.

7. Czochralski Growth. The principal technique applicable to growth of fluorides is the Czochralski/topseeded solution method which is characterized by seeded, unconfined crystallization in an open crucible but maintained in a rigorously pure oxygen-free and water-free argon atmosphere. The furnace hardware and the computer control system are continually scrutinized, studied and where possible updated,to conform to ever increasing demands for crystal quality, purity and process relability.

The furnace system consists of a water cooled stainless steel chamber evacuated by a 4 inch oil diffusion pump as shown in Fig. 4. Almost all of the various chamber subassemblies are of standard flanged-elastomer 'o'-ring sealed design. The picket fence style resistance heater and crucible supports are all made from graphite. Molybdenum sheet is used for thermal shielding; no other insulation is used. Other parts are stainless steel; the windows are of quartz. Generally platinum crucibles are used. Single crystal seeds, or the platinum wire used in their absence, are attached to the end of a water-cooled seed rod. This is sealed through the top of the furnace using a double 'o' ring gland with a guard vacuum allowing leak free rotation and translation. The furnace chamber is divided by a horizontal gate valve into two subchambers, the lower one housing the melt. This arrangement allows, without shutdown and exposure to the atmosphere, critical cleaning of the melt surface in a preliminary crystal growth experiment followed by removal of the contaminated material to the upper chamber and introduction of a clean single crystal seed. The furnace is prepared for crystal growth by thorough cleaning of the evaporated residue from the previous run,followed by outgassing in high vacuum up to temperatures of 1300°C. After this treatment, the feed is loaded and the

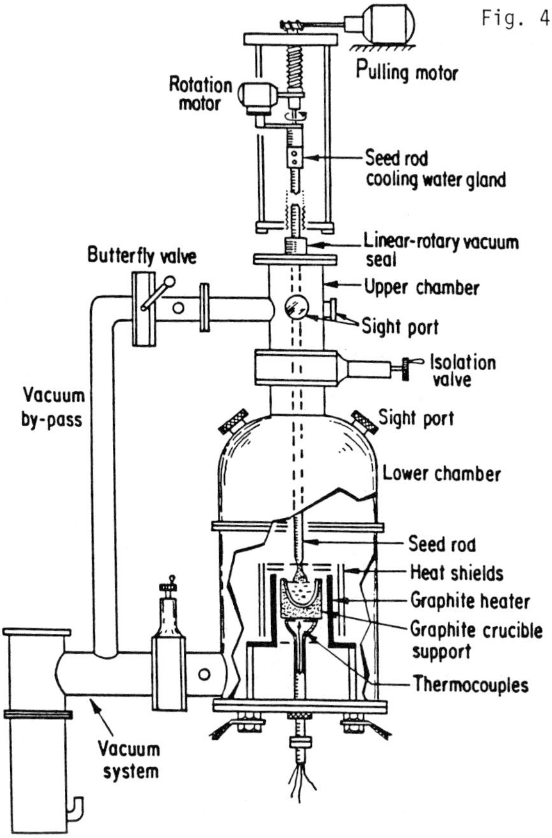

Fig. 4. Fluoride Pulling Furnace

furnace is re-outgasses, but at lower temperatures, to avoid vaporization of the feed. Crystal growth is carried out in high purity Ar which is admitted to the furnace after the outgassing process is judged to be complete.

By processing under such careful control, it is possible to grow single grain boules with a mosaic spread of less than 2" arc over many cubic centimeters volume.

Control of the crystal growth process is executed by a dedicated Data General Micronova computer. The control algorithm, using inputs from a thermocouple, Hall-effect power-measuring device and a He-Ne laser-based optical diameter sensing system, as shown in Fig. 5, is based on a systems analysis derived from modern control theory.

With the crucible open to the whole working volume of the furnace, difficulty can arise when evaporation from the melt is excessively great, or when the oxidation state of a dopant is not stable. The latter problem can sometimes be approached by introducing a carefully controlled partial pressure of a reducing agent into the growth atmosphere.

While simple and rapid-melt experiments in the hydrofluorinator or the zone-refiner can yield small samples of fluorides for cursory fluorescence measurements, determination of whether or not a material has potential for

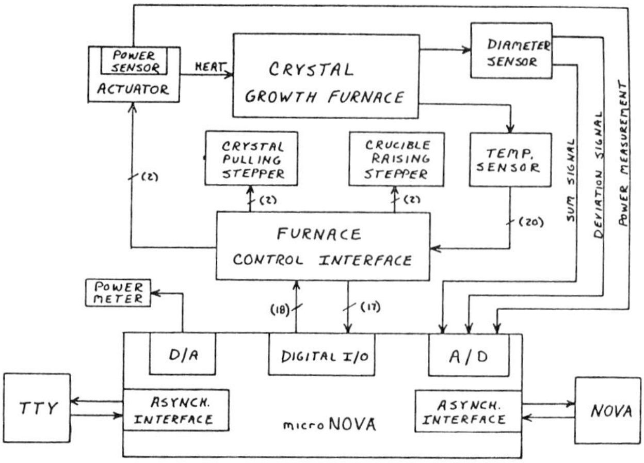

Fig. 5. Function Block Diagram of the Microprocessor Controlled Crytal Growth System

use as a laser requires a more serious study of the growth process, usually by the Czochralski technique. During such a growth experiment important qualitative visual observations can be made. When the solid feed is melted in argon we observe an approximate melting point. The melt must be clear; cloudiness may indicate presence of a second phase, possibly due to an impurity. At this stage it will begin to become apparent if any of the components has an excessively high vapor pressure; evaporation and condensation on the cold seed rod can be observed. Crystallization is nucleated on a platinum seed, attached to the water-cooled seed rod, by lowering the melt temperature. When the wire reaches the temperature of the liquidus a disc of solid phase will begin to form. If the crystalline material is not transparent this indicates that two or more phases are crystallizing simultaneously. A clear disc indicates single phase crystallization. In either case, using the closed loop diameter control mode, a cylinder is crystallized and eventually disconnected from the melt and cooled to room temperature. During the growth-cool down cycle it is often possible to observe decomposition of solid solutions (opacification of a formerly clear material) and destructive phase transitions.

These numerous significant facts are learned through close visual observation of the first crystal growth experiments. X-ray diffraction analysis of the resulting specimen will show what phase it is, i.e. whether it is the desired material. If the identity and number of phases present varies along the length of the cylinder, other information about the phase diagram may be inferred.

All of these things must be experimentally evaluated. The most direct procedure is to prepare a melt of the stoichiometric composition and proceed directly to grow a crystal, observing various phenomena germane to successful crystal growth. According to such a mode of operation, the component fluorides are first melted together in the hydrofluorinator. Then the prepared feed is analysed by x-ray diffraction to determine what phases are present.

An initially two-phase region giving way to a single-phase region suggests incongruent melting, while the appearance of a two-phase region after crystallization of a single-phase delineates a two-phase boundary and defines the maximum yield. Not all of this information, however, is necessarily obtained in a single crystallization experiment. With certain crystal growth problems defined, crystal growth research may then be directed toward their solution.

Summary. These techniaues have been applied successfullv to the arowth of a number of fluoride laser host systems. These include congruentlv melting materials such as BaY_2F_8 and incongruently melting ones such as $LiYF_4$ (YLF) and its rare earth isomorphs, and, more recently, the very incongruent material $LiBaF_3$, which appears to be an interesting new host for divalent rare earths. Experimental work on elpasolites and other complex fluorides is in progress.

Growth of $Co:MgF_2$ and $Ti:Al_2O_3$ Crystals for Solid State Laser Applications

F. Schmid and C.P. Khattak
Crystal Systems, Inc., 35 Congress Street, Salem, MA 01970, USA

Large, 6.5 cm diameter, crystals of $Co:MgF_2$ and $Ti:Al_2O_3$ have been grown by the Heat Exchanger Method. The $Co:MgF_2$ crystals are of laser quality and show uniform Co distribution. Power conversion efficiency of 32.5% was measured for this tunable laser material (1.5 to 2.3 µm). Fine scattering centers have been observed in $Ti:Al_2O_3$ crystals. A quantum efficiency of 62% was demonstrated with continuous tunable laser operation in the 660 to 986 µm wavelength range.

1. Introduction

In most lasers, the energy is released via stimulated emission in the form of photons. Recently a new class of "vibronic" solid state laser material has emerged, in which the stimulated emission is coupled with the photons as well as phonons. Partitioning of the total energy between the two results in broad tunability of the laser output. Even though the tunable operation of lasers using vibronic impurity elements was recognized in early work [1], there has been renewed interest in solid state tunable lasers recently. For example, Allied Corporation is now commercially offering lasers using alexandrite, Cr^{3+} doped $BeAl_2O_4$, which permits tunable laser operation in the 700-820 nm wavelength region [2,3] . The rediscovery of CW operation in divalent doped MgF_2 [4,5] has revived interest in solid state tunable lasers and has led to the discovery of new materials [6-8].

This paper describes growth of large crystals of two promising tunable laser materials - $Co:MgF_2$ and $Ti:Al_2O_3$. Of the divalent doped MgF_2 crystals, $Co^{2+}:MgF_2$ is most promising because the Ni^{2+} and V^{2+} materials exhibit excited state absorptions. Growth of these crystals has been demonstrated by a vertical gradient freeze technique [9]. The present paper describes the adaption of the Heat Exchanger Method (HEM™) [10] for the growth of $Co:MgF_2$ crystals up to 6.5 cm diameter size. Large, 6.5 cm diameter crystals of Ti^{3+} doped sapphire ($Ti:Al_2O_3$) have also been grown by HEM. In this material it is important to maintain the titanium in the trivalent state: Ti^{4+} is not suitable for lasing action because it forms scattering centers in the crystal leading to loss of power conversion efficiency. High quality $Ti:Al_2O_3$ crystals have been grown; however, the material shows scattering centers under intense light or laser illumination. Laser data has been obtained at MIT Lincoln Laboratory on HEM grown $Ti:Al_2O_3$ crystals.

2. Heat Exchanger Method (HEM)

The HEM is in commercial production for large sapphire and silicon ingots [11,12]. The salient features of the process are shown in Fig. 1. The crucible with the seed positioned at the bottom is loaded with charge and placed on top of the heat exchanger. After evacuation, heat is supplied by the graphite heater. The HEM furnace contains no moving parts; hence, it is easy to introduce a controlled atmosphere, such as argon gas, in the furnace

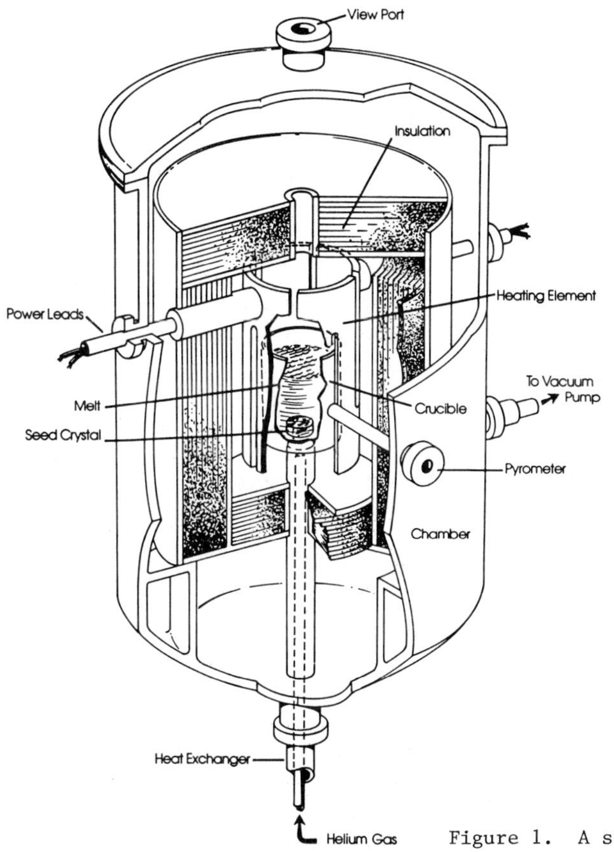

Figure 1. A schematic of HEM furnace

chamber. The charge is melted and the seed is prevented from melting by forcing gaseous helium through the heat exchanger. After sufficient meltback of the seed is achieved, growth is started by increasing the flow of helium and thereby decreasing the heat exchanger temperature. The liquid gradients are controlled by the furnace temperature while the gradients in the solid by the heat exchanger. HEM, therefore, has the unique feature of independent control of solid and liquid gradients with no moving parts. This feature is significant, in that in the heat zone no gradients are built in; different size crystals can, therefore, be grown in the same HEM furnace by changing the temperatures and thereby the gradients. An additional advantage is the flexibility to alter the gradients during the crystal growth cycle.

A distinguishing feature of the HEM, as compared to Czochralski or top-seeded processes, is that the solid-liquid interface is submerged beneath the surface and is surrounded by the melt. The thermal and mechanical perturbations are, therefore, damped out by the surrounding molten mass resulting in uniform temperature gradients at the interface. These features of HEM produce uniform growth and the only sapphire which is free of light scatter [11].

In the case of sapphire and silicon, it has been demonstrated that once crystal growth parameters are established, large crystals can be grown.

3. Growth of Co:MgF_2 Crystals

Large, optical quality, undoped MgF_2 crystals are routinely grown from the melt in open crucibles; however, this method cannot be used for growth of doped crystals because of the volatile nature of the transition-metal-dopant fluoride in comparison to the host material at its melting point. Laser quality doped crystals have been grown in self-sealing graphite crucibles by a vertical gradient freeze technique [9].

Initial experiments with HEM were for growth of undoped MgF_2 crystals. A 2.5 cm diameter graphite crucible with a tapered cover was fabricated. The internal surfaces of the crucible were cleaned with sandpaper to remove particles of graphite followed by polishing using alumina powder. After removal of loose graphite particles, the crucible was thoroughly washed with soap and water, rinsed with alcohol and dried. The crucible was loaded with optical grade MgF_2 chips and placed on the heat exchanger in the HEM furnace. After evacuation of the chamber, heat was applied to the heating element. At about 1100° the furnace was backfilled with argon gas to atmospheric pressure. After meltdown, the furnace temperature was increased to achieve superheat and stabilize the melt. Crystal growth was initiated by increasing the flow

Figure 2. A 2.5 cm diameter MgF_2 crystal grown by HEM

Figure 3. A 4.4 cm diameter, 3.5 cm high Co:MgF_2 grown for laser application

of helium through the heat exchanger and decreasing the furnace temperature. After solidification, the heat exchanger temperature was increased to anneal the crystal below its melt point, and the crystal was cooled to room temperature at a controlled rate.

Figure 2 shows the 2.5 cm diameter undoped MgF_2 crystal grown by HEM. Even though no seed was used, the entire material grew as a single crystal. The growth direction was oriented about 50° to the c-axis, which is close to the [322] direction. A polished slab sectioned perpendicular to the c-axis was examined under cross-polarizers and intense light illumination. No bubbles, precipitates and grain boundaries were seen in the sample.

Similar procedures were used for seeded growth experiments except that during the meltdown stage, a minimal helium flow was forced through the heat exchanger. For doped-crystal growth, CoF_2 powder was added to the charge in the crucible. A 4.4 cm diameter, 3.5 cm high Co:MgF_2 crystal grown by HEM is shown in Fig. 3. The as-grown crystal, to the naked eye, was dark and unifom in color.

The growth axis of the crystal was approximately 30° from the c-axis. After orientation 6 mm thick c-axis slabs were sectioned and polished for examination. Absorption spectroscopy showed a uniform 0.9 + 0.05 wt% cobalt doping in all the slabs. Examination for scattering under a laser beam with a detectability limit of 0.001 cm^{-1} showed no scattering centers. The

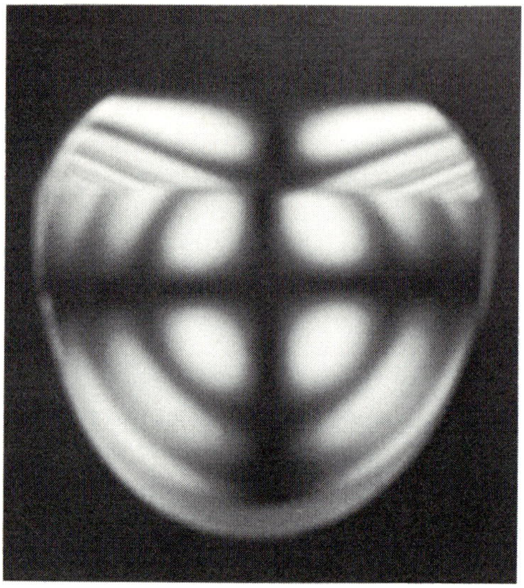

Figure 4. Examination of a c-axis Co:MgF slab under cross-polarizers

slabs under cross-polarizers showed no distortion of the fringe patterns (Fig. 4).

After establishing the growth of high quality Co:MgF_2 crystals for laser applications a 6.5 cm diameter, 4.5 cm high crystal was grown by HEM. Characterization of this material showed that this crystal was of similar quality to the smaller crystals.

4. Growth of Ti^{3+}:Al_2O_3 Crystals

Sapphire, the host crystal, is grown free of light scatter using HEM [11]. These crystals are grown in molybdenum crucibles under vacuum. Initially, a 6.5 cm diameter molybdenum crucible was loaded with a sapphire seed, Ti_2O_3 powder and sapphire crackle and a crystal was grown using routine procedures. The pink-colored crystal showed that considerable loss of Ti:Al_2O_3 had occurred. Under intense light illumination, small bubbles were seen throughout the boule as scattering centers. It is postulated that under vacuum conditions, there is volatilization and the product is entrapped in the growing crystals causing bubble formation.

Experiments were, therefore, carried out where the HEM furnace was back-filled with argon gas prior to meltdown. A 6.5 cm diameter, 6 cm high Ti:Al_2O_3 crystal grown under these conditions is shown in Fig. 5. The volatilization problem was suppressed. This crystal was examined for scattering centers using intense white-light illumination as well as a laser beam. Fine scattering centers were dispersed in localized areas of the crystal. Visual observation showed a variation in color along the growth

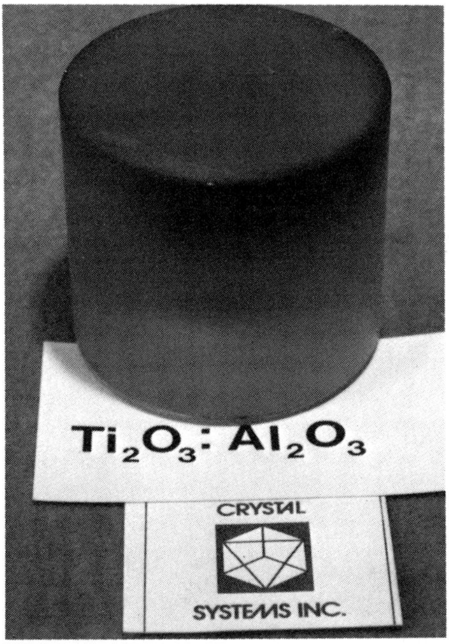

Fig. 5. A 6.5 diameter, 6 cm high $Ti{:}Al_2O_3$ crystal grown by HEM

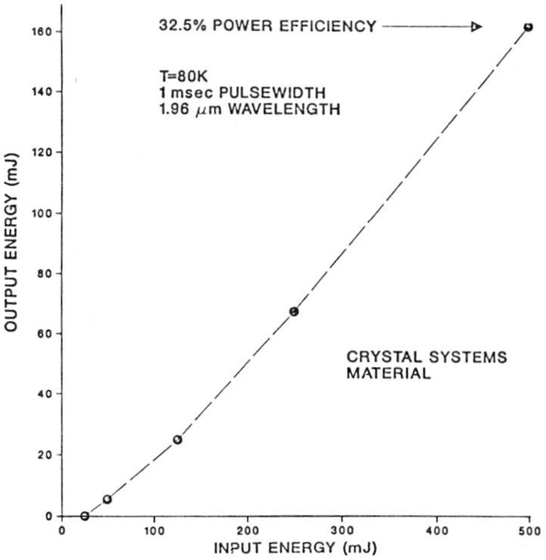

Fig. 6. The input vs. output energy for a $Co{:}MgF_2$ laser rod optically pumped by a Nd:YALO laser

axis. The material was darker as growth progressed, indicating that Ti^{3+} has a segregation coefficient which is less than unity. Optical absorption measurements were carried out to estimate the Ti doping concentration. The data showed a Ti variation from 0.053 to 0.13 atomic percent using a reference sample containing 0.1 atomic percent.

5. Laser Performance

The fluorescence of $Co{:}MgF_2$ under pulsed-pumping conditions shows a very wide tunability range and can be tuned continuously from 1.5 to 2.3 μm [6]. One of the slabs of HEM grown $Co{:}MgF_2$ crystal was fabricated into a Brewster-angle rod. This laser crystal, 6 mm x 6 mm x 47 mm was optically pumped by a 1.34 μm Nd:YALO laser in a two mirror cavity system at MIT Lincoln Laboratory. The power conversion efficiency of 32.5% was measured for this laser material (Figure 6).

A qualitative analysis of incident light absorption of a $Ti{:}Al_2O_3$ sample showed that nearly all the light was absorbed in the laser rod indicating that the dopant level was high enough. Based upon this data, a 6 mm diameter, 2 cm long laser rod was fabricated. After polishing and AR coating the ends, this rod was tested for lasing action at MIT Lincoln Laboratory. The rod was placed in an optical cavity consisting of 25 cm radius mirrors spaced about 25 cm apart at room temperature. The sample was optically pumped by a

Rhodamine 6G dye laser operating at 587 nm. A quantum efficiency of 62% was measured with continuous tunable laser operation in the 660 to 986 nm wavelength range.

6. Conclusions

Tunable laser operation, with a wide tunability range, has been demonstrated for $Co:MgF_2$ and $Ti:Al_2O_3$ crystals. In both cases, 6.5 cm diameter crystals have been grown; these are the largest sizes ever grown in these materials. High laser quality $Co:MgF_2$ crystals with minimal scattering centers and with uniform Co distribution have been fabricated. In the case of $Ti:Al_2O_3$, fine scattering centers have been observed. While this material has been suitable for testing purposes, higher quality is necessary to demonstrate the full potential of this material.

7. Acknowledgment

The authors wish to thank Dr. Peter Moulton of MIT Lincoln Laboratory for technical discussions and laser performance data. This material is based upon work supported by the National Science Foundation under grant numbers DAR-8018284 and DMR-8260712.

8. References

1. L. F. Johnson, R. E. Dietz and H. J. Guggenheim, Phy. Rev. Lett. 11 318 (1963)

2. J. C. Walling, O. G. Peterson, H. P. Jenssen, R. C. Morris and E. W. O'Dell, IEEE J. Quant. Electron. QE-16, 1302 (1980)

3. J. C. Walling, Laser Focus 18 (2) 45 (1982)

4. P. F. Moulton, A. Mooradian and T. B. Reed, Optics Lett. 3, 164 (1978)

5. P. F. Moulton and A.Mooradian, Appl. Phys. Lett. 35, 838 (1979)

6. P. F. Moulton, IEEE J. Quant. Electron. QE-18, 1185 (1982)

7. P. F. Moulton, Laser Focus 19 (5) 83 (1983)

8. B. Struve, G. Huber, V. V. Laptev, I. A. Shcherbakov and E. V. Zharikov, Appl. Phys. B30, 117 (1983)

9. T. B. Reed, R. E. Fahey and P. F. Moulton, J. Crystal Growth 42, 569 (1977)

10. F. Schmid and D. Viechnicki, J. Am. Ceram. Soc. 53, 528 (1970)

11. F. Schmid and C. P. Khattak, Laser Focus/Electro-Optics 19 (9) 147 (1983)

12. F. Schmid and C. P. Khattak, Optical Spectra 15 65 (1981)

The Laser-Heated Pedestal Growth Method: A Powerful Tool in the Search for New High Performance Laser Crystals

R.S. Feigelson
Center for Materials Research, Stanford University, Stanford, CA 94305, USA

1. Introduction

Various predictive methodologies can be used to help identify new materials for improved or advanced laser device applications. The predictions, however, are of questionable reliability and therefore have to be verified by 1) synthesizing the compounds thought to be most promising, 2) preparing them in the form of high quality single crystals, and 3) measuring their properties. The route from the prediction stage to the point where a device-quality crystal with appropriate performance characteristics can be identified is a long and tortuous one, which the crystal growth phase being the most expensive and lengthy part. The difficulty, cost and years of development time required to produce these crystals has definitely hindered the development of many important devices and systems requiring high performance crystals.

A promising new approach to help solve the crystal growth problem involves growing small crystal fibers 200-1000 μm diameter by the laser heated pedestal growth method (LHPG). The simplicity of the method is illustrated with the photograph in Fig. 1 which shows an actual fiber crystal being grown. Because the sample sizes are small,the source material is easy to prepare, elaborate furnaces and expensive crucibles are not required, and the growth rates are much faster than are permissible in the growth of larger crystals (mm/min compared to mm/hr). Thermal stresses are minimized as are crystallographic defects and very high melting oxides and fluorides can easily be melted by the laser heating method. Due to these factors this technology provides the fastest, most efficient and inexpensive approach to the material survey process and will speed the development of new high-performance laser crystals.

Once a promising laser host and dopant combination has been identified from the fiber crystal growth phase, larger crystals of this material can be grown and evaluated.

Fig. 1 A photograph showing the inherent simplicity of the LHPG method when used for the growth of small diameter crystals

2. Historical Background

GASSON AND COCKAYNE [1] were the first to use laser heating for crystal growth. They prepared relatively large single crystal boules of high melting oxides such as Al_2O_3, Y_2O_3, $MgAl_2O_4$ and Nd_2O_3 by the float-zone method. The large thermal gradients coupled with large crystal dimensions led, however, to numerous difficulties.

Several years later HAGGERTY [2] reported on the development of a four beam laser-heated float-zone system specifically designed for the preparation of single crystal fibers. He grew fibers of Al_2O_3, Y_2O_3, TiC and TiB_2, principally for evaluation of their mechanical properties. Haggerty used a 200 W CO_2 laser in conjunction with a spherical reflector and a furnace chamber which permitted a controlled atmosphere including vacuum. He grew fibers oriented in several directions and explored various furnace atmospheres, growth orientations, growth rates, beam energy densities and angles of incidence, and radiation shielding.

A two beam laser heated pedestal growth system similar to Haggerty's was developed by BURRUS AND STONE [3] in 1975 to grow single crystal Nd:YAG fiber lasers as small as 50 μm in diameter. These crystals achieved room temperature CW laser operation in 0.1-1 cm long samples.

STONE et al. [4] grew oriented Nd:YAG fibers by the same method and studied their surface morphology.

STONE AND BURRUS [5] also grew Nd^{+3} doped Y_2O_3 fibers which achieved laser operation at room temperature at 1.07 and 1.35 μm. Starting with high purity Y_2O_3 and Nd_2O_3 powder compacted into a cylindrical bar, fibers 60-120 μm in diameter and free of visible scattering defects were grown at rates of 3mm/min. Unlike YAG and Al_2O_3 fibers, these crystals had relatively rough surfaces and a milky appearance resulting from Y_2O_3 vapor recondensing on the surface. This coating could be readily removed in hot dilute HCl.

BURRUS AND COLDREN [6] prepared sapphire-clad ruby fibers by the same method. After first growing a ruby crystal,they then melted and regrew the surface of the fiber with the laser. In this way the Cr concentration was reduced in the surface layer by a factor of 100 or more by selective evaporation.

About five years ago a broad program was started at Stanford University to study the growth of fiber crystals, their properties and devices which might incorporate them. While interest in advanced devices provided the major driving force for this work on single crystal fiber growth, there were also many other interesting and challenging problems related to the growth of these small crystals, including understanding their solidification behavior, the formation and morphology of ferroic domains and eutectic phases, and a study of crystal defects. It was also apparent from the beginning that one of the most powerful applications of the laser-heated fiber growth method would be for the growth of new materials in single crystal form for property evaluation, particularly laser hosts suitably doped with activator ions. These factors have led to a broad program involving a large number of materials for a variety of optical and electronic applications.

The laser-heated pedestal growth method was chosen as the principal growth technique for the above program because of its great versatility [7]. In addition to the high growth speeds made possible by the steep temperature gradients, the technique is containerless and contamination and stress problems are minimized. The crystals can easily be doped with a variety of ions. Very high temperatures can be achieved with laser heating and the heating and cooling rates are very rapid. The shape of the molten zone, melt convection, macroscopic crystal defects and, in transparent systems, the interface shape can usually be observed directly while the crystal is growing. Since the sample size is small, material costs are low. This is important when expensive materials such as scandium compounds are being studied. Post-growth laser annealing can be used to alter purity, composition and surface structure. The fast growth rates should permit increased dopant incorporation and may lead to the preparation of a number of new metastable materials.

3. The Pedestal Growth Method

The Stanford fiber growth system, shown schematically in Fig. 2, was based on the one developed by Burrus and Stone. A 50 watt CO_2 laser, operating at 10.6 μm, is split into two beams which are focused onto the end of a source rod containing the desired compound and dopant (if desired). This is illustrated in Fig. 3a and 3b where the first stages of melting are shown. The fiber crystal is then withdrawn from the molten zone as shown in Fig. 3c and 3d. This system in its present configuration can produce fibers 3-1700 μm in diameter in lengths of up to 20 cm. The system contains a sealed growth chamber, and both the fiber and source rod can be rotated. The growth speed can be varied in the range of 0.1-100 mm/min. The laser and growth station are mounted on a vibration-damped table and the growing crystal and molten zone are observed through a binocular microscope with variable magnification up to 70x.

The amount of power necessary to melt the sample is proportional to the source rod diameter. Oxide materials absorb 10.6 μm radiation very efficiently and a nominal 50 watt CO_2 laser is ample to melt all oxide

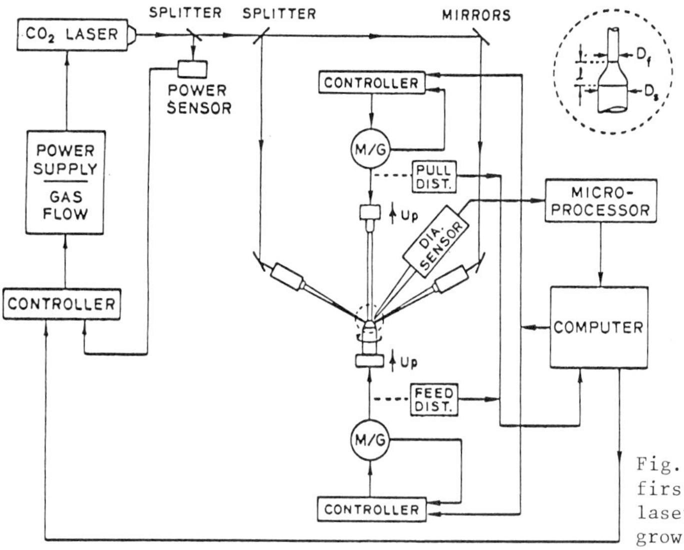

Fig. 2 Schematic of the first generation two beam laser-heated pedestal growth apparatus

a

b

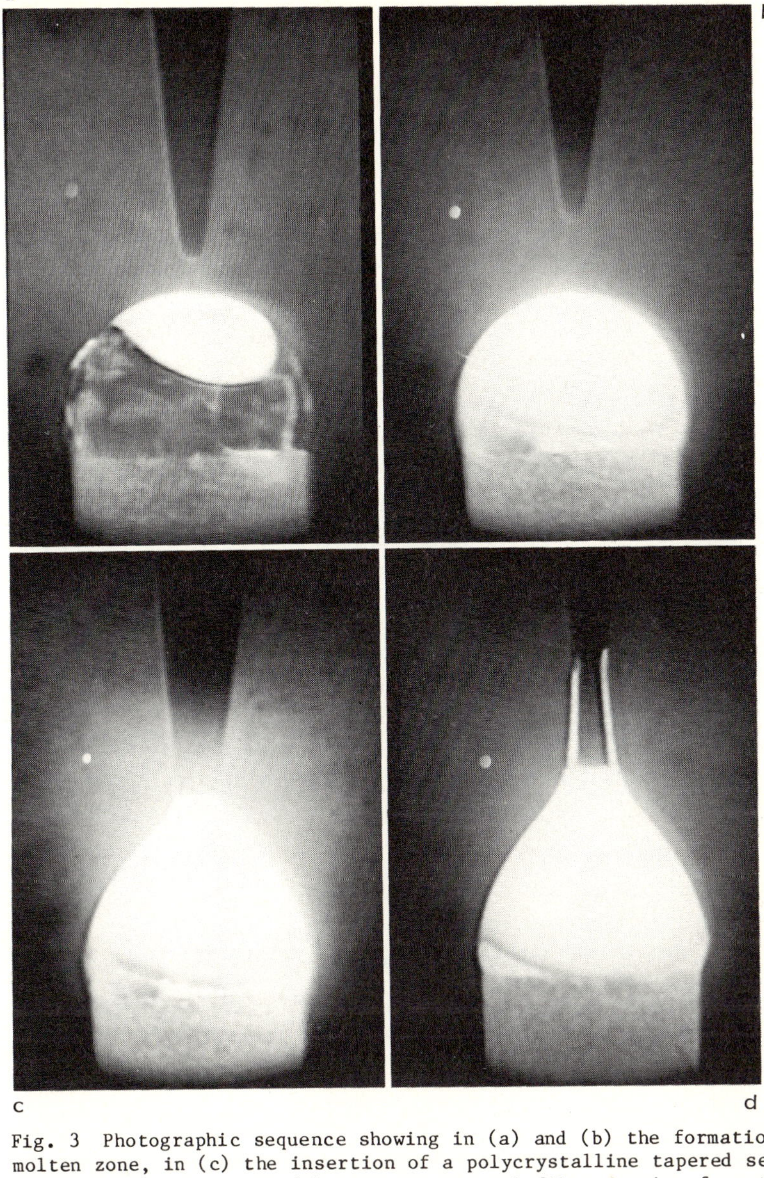

c

d

Fig. 3 Photographic sequence showing in (a) and (b) the formation of the molten zone, in (c) the insertion of a polycrystalline tapered seed to initiate growth, and in (d) a single crystal fiber growing from the polyseed

compounds less than 1 mm in diameter as shown in Fig. 4 for three typical oxide compounds, Al_2O_3, $LiNbO_3$, and $Y_3Al_5O_{12}$ (YAG). When the source rod is less than 100 μm, the power required to form and maintain a stable molten zone is so small that it is difficult to maintain laser stability. The use of a power attenuator allows the laser to operate in a stable mode at mid-power levels, even when less than 1 watt is required for melting.

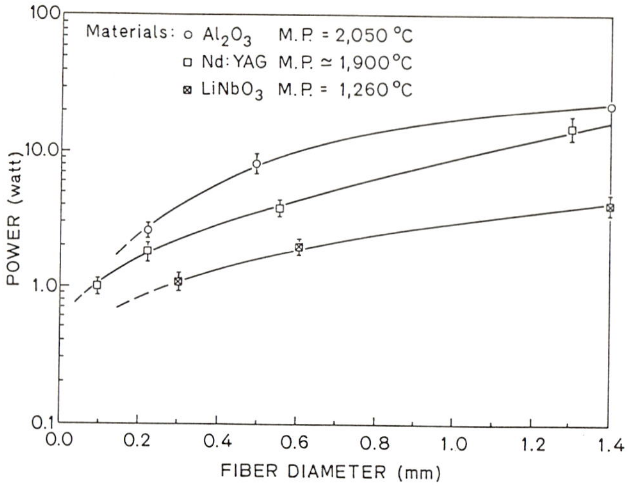

Fig. 4 The laser power necessary to form a molten zone in Al_2O_3, Nd:YAG and $LiNbO_3$ as a function of source rod diameter

Fluorides also absorb the 10.6 μm effectively and can be easily melted and grown by the LHPG method.

When the fiber starts to grow, the source rod is fed into the laser beam so as to maintain a constant melt volume. In float zone growth the diameter of the crystal and feedstock are the same. In the pedestal growth method the fiber diameter is typically between 1/2 to 1/4 the diameter of the source rod and so the ratio of fiber to source rod pull and push rates have to be adjusted accordingly. In the photographs shown in Fig. 3 the fiber is being pulled upward, but it is common to pull fibers downward as well. The angle of the beam, beam diameter and shape, and laser stability are all important factors in determining the shape and stability of the molten zone and the quality of the crystals produced.

The two-beam system introduces large and sometimes undesirable radial temperature gradients into the molten zone depending upon the thermal properties of the material, the source rod size and beam diameter (typically 100 - 500 μm). A spherical reflector can reduce these gradients somewhat. The four-beam system used by Haggerty reduces the problem as does rotating the fiber and source rod. This problem is particularly important when large diameter source rods are used. A new optical lens system using parabolic mirrors has recently been designed, constructed, and evaluated by FEJER et al. [8]. This patented optical system [9] gives a circular beam and a uniform hot zone and seems ideally suited for use with the LHPG method.

Unlike the glass fiber drawing process, where the high viscosity of the melt holds the hot zone in alignment, the melts of most crystalline materials have a low viscosity and the fiber is free to move if disturbed. Single crystal fiber growth therefore requires a vibration-free environment and one in which air convection is minimized. A sealed chamber therefore is very important to the achievement of uniform fiber diameters and also allows the ambient atmosphere to be varied from reactive to inert. In some cases a vacuum chamber may be necessary or desirable. For good diameter control therefore, particularly with the growth of very small diameter fibers, you must have in addition to thermal symmetry, excellent mechanical and thermal stability.

A video monitoring and recording system has been incorporated into the LHPG system (Fig. 5). It provides quantitative data which can be used to evaluate growth phenomena in real-time. Replay of video tapes permits a more in-depth evaluation of changes in zone and interface shape, defect formation, melt convection, etc. It could also be used for diameter control.

Source rods for fiber growth are very easy to prepare. Rods can be cut from single crystal or polycrystalline samples. Square cross-section bars are adequate in most instances but round rods are preferable. To help achieve uniform diameter fibers,the diameter of the source rod should be very uniform, particularly when automatic diameter control is not available.

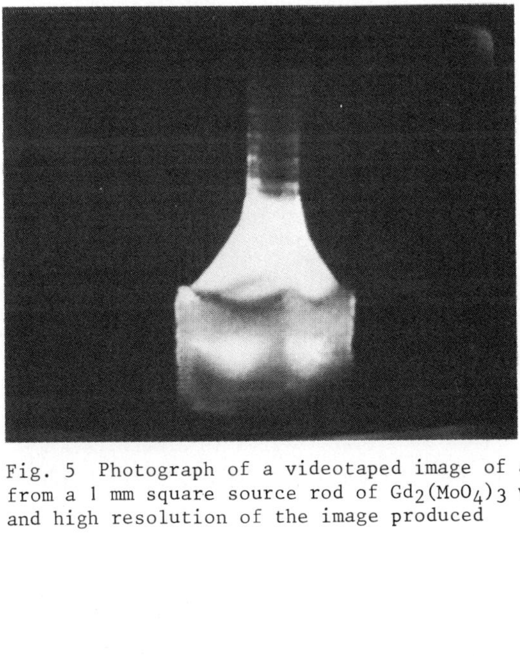

Fig. 5 Photograph of a videotaped image of a fiber crystal being grown from a 1 mm square source rod of $Gd_2(MoO_4)_3$ which illustrates the clarity and high resolution of the image produced

Source rods are usually 0.5-1 mm in diameter and smaller diameter fibers are obtained by repetitive growths using the first-grown fiber as the next source rod and so on. Using a 3:1 reduction ratio, a 1 mm starting source rod can be turned into a 37 μm diameter fiber in three growth runs.

Doping can be accomplished by several methods. If the dopant has a low vapor pressure, a given quantity of dopant can be coated on the surface of the source rod, either by evaporation, or by painting or spraying on a liquid suspension containing the dopant. The dopant will then dissolve into the molten zone continuously as the source rod is melted. Dopant can also be mixed with the starting powders and then hot pressed or cold pressed and sintered into dense polycrystalline ceramics. Other methods are also possible.

To initiate the growth of a single-crystal fiber, a wire, tapered polycrystalline ceramic (as shown in Fig. 3) or single crystal can be used for seeding. Single-crystal growth is very easy to achieve with any of these methods. In fact, it is difficult to inhibit single-crystal growth. Crystals can be grown along specific crystallographic directions by using oriented seed crystals.

4. Characterization

Fibers can be characterized with regard to orientation, crystal defects and their chemical and physical properties by standard characterization techniques such as x-ray diffraction, optical microscopy, etching, etc., although their small size makes handling and adaption to conventional methods a more difficult task.

Optical microscopy to date has been one of the most important techniques for studying ferroelectric domains, eutectic morphology and macroscopic defects such as voids, precipitates and optical inhomogeneities in fibers.

Techniques for evaluating laser crystals in fiber form have been established [13].

5. The Growth of Laser Crystals

Some of the oxide and fluoride crystals grown at Stanford by the LHPG method are given in Table I, along with their respective melting temperatures. Two of these materials, Nd^{+3}:YAG, $Ti^{+3}:Al_2O_3$ are important laser crystals. $Nd^{+3}:CaSc_2O_4$ was used as a model system to demonstrate how effective the method is for rapidly surveying new materials for laser applications. The growth of these three materials will be discussed below.

Table I: Typical Single Crystals Grown by the LHPG Method

COMPOUND	MELTING TEMPERATURE °C	COMPOUND	MELTING TEMPERATURE °C
$Y_3Al_5O_{12}$	1940	$CaSc_2O_4$	2110
$Y_3Al_5O_{12}:Nd^{+3}$		$CaSc_2O_4:Nd^{+3}$	
Al_2O_3	2045	$Y_3Fe_5O_{12}$	1555
$Al_2O_3:Ti^{+3}$		$Y_2O_3:Eu$	2410
$Al_2O_3:Cr^{+3}$		Nb_2O_5	1495
$LiNbO_3$	1260	$ScTaO_4$	2300
$LiNbO_3:Nd^{+3}$		$ScNbO_4$	2100
$LiGeO_3$	1170	BaF_2	1280
$Gd_2(MoO_4)_3$	1157	CaF_2	1360
		NaF-LiF	652

5a. $\underline{CaSc_2O_4}$

The compounds calcium and strontium scandate ($CaSc_2O_4$ and $SrSc_2O_4$) were first synthesized by CARTER AND FEIGELSON [10] in 1964. They determined their crystal structure by very careful powder x-ray diffraction techniques. The crystal structure analysis would have been greatly facilitated had single crystals of these compounds been available. However, the growth of these compounds would have been prohibitively expensive and difficult because 1) they have very high melting temperatures (in excess of 2000°C), 2) Sc_2O_3 was very expensive, 3) costly, nonreactive crucibles would have been required to contain the melt, and 4) at that time convenient techniques for growing crystals of materials which melt at temperatures greater than 2000°C had not been developed. Solution and vapor-growth techniques, which typically allow lower temperatures, would have required a much lengthier period to develop a process capable of producing even a small crystal sample.

During the early phases of the LHPG program KWAY AND FEIGELSON [11] looked for a model material which would be representative of the growth of a new high-temperature material for evaluation as a solid state laser host. They found a 20-year old hot pressed disc of $CaSc_2O_4$ from the earlier synthesis work of CARTER AND FEIGELSON [10] which seemed ideal for this purpose. The disc was cut into a rod with a 1 mm square cross section and 2 cm in length. In three hours, including set-up time, the first single crystal of $CaSc_2O_4$ was produced. This crystal, shown in Fig. 6, was grown using a seed cut from the same hot pressed disc. This seed was tapered to a point. The fiber crystal that grew from it had a

Fig. 6 A single crystal fiber of $CaSc_2O_4$ grown along the a-axis by the LHPG method

random orientation. Using x-ray Laue techniques, the major crystallographic directions were identified and an oriented crystal was then grown by mounting the as-grown crystal at an appropriate angle to the growth direction as shown in Fig. 7.

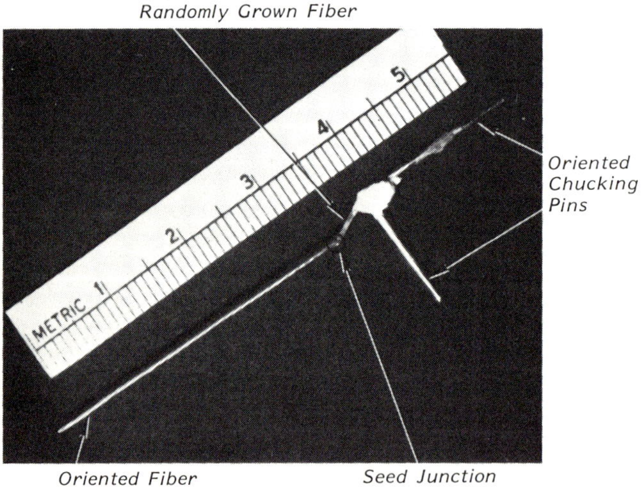

Fig. 7 Photograph illustrating the method by which oriented fiber crystals are grown

The next challenge was to see if this compound could be made into a laser crystal by the incorporation of a suitable dopant. BAGDASAROV et al. [12] reported on the stimulated emission properties of 5 mm diameter $Nd^{3+}:CaSc_2O_4$ crystals which were presumably grown from the melt, although no details were given. Neodymium doped $CaSc_2O_4$ fiber crystals were prepared by the LHPG technique by coating the surface of our hot pressed source rod with a suspension of Nd_2O_3 in toluene by a simple brushing technique. During growth the Nd_2O_3 dissolved into the molten zone continuously as the source rod was consumed. This dopant technique worked relatively well although a more sophisticated technique such as vacuum sputtering has been used to achieve better control of the dopant concentration and uniformity.

DIGONNET et al. [13] found that the $Nd^{3+}:CaSc_2O_4$ fibers produced by this technique lased at 1.075 μm and had a fluorescence spectra at 300° K as shown in Fig. 8. This data compares well with that of BAGDASOROV et al. [12]. The amount of material consumed in this project was minimal and the growth process, crystal orienting and doping steps took very little time compared with other methods.

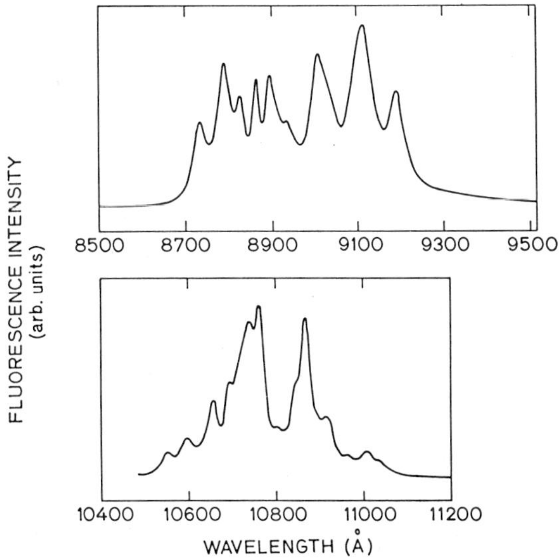

Fig. 8 The fluorescence spectra of a Nd^{3+} doped $CaSc_2O_4$ crystal from DIGONNET et al. [13]

5b. Nd^{3+}:YAG

One of the long term problems associated with the growth of large Nd^{3+}:YAG laser crystals is facetting at the growth interface. This leads to the segregation of Nd at the facets, which in turn gives rise to lattice strain and optical index inhomogeneties in the core region of the crystal. A program was recently started to find out whether this problem also occurs during the growth of Nd doped YAG fiber crystals, and if so, to determine the relationship between fiber diameter and the radial segregation behavior of the Nd dopant. Both the thermal conditions and the growth rates used in the LHPG method are different to those found in Czochralski growth, the method used to grow large Nd:YAG crystals, and, therefore, a change in the segregation behavior might be expected.

Electron microprobe analysis is being used to determine both the radial and axial variations in Nd concentration in fiber crystals produced under various growth conditions. In Fig. 9 a radial microprobe scan is shown for two fibers, one grown by BURRUS AND STONE [3] and a second grown by O'MEARA et al. [14] at Stanford. A marked increase in Nd concentration was found in the center of the Burrus and Stone crystal, even though it had a very small diameter. The crystal grown by O'Meara et al., was close to the same diameter, but showed a greater uniformity in dopant concentration. The segregation behavior of Nd in YAG seems therefore to be influenced by the growth conditions rather than diameter. More extensive work will be carried out to determine the relationship between growth parameters, diameter and Nd concentration profiles.

5c. $Ti^{3+}:Al_2O_3$

An important new crystal for tunable solid state vibronic lasers is Ti^{3+} doped sapphire. These crystals have been produced in bulk form with

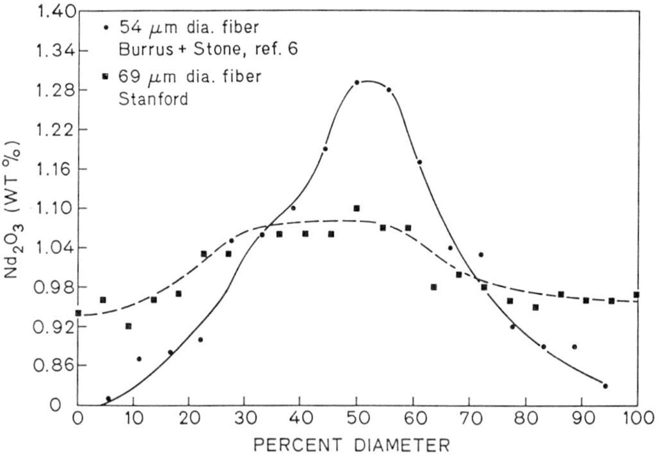

Fig. 9 Electron microprobe analysis of two Nd:YAG fiber laser crystals, one from BURRUS AND STONE [3], and the other grown by O'MEARA et al. [14], showing radial variations in Nd concentration

success, but questions still remain concerning the maximum solubility limit of trivalent Ti in Al_2O_3, and how to control the precipitation of TiO_2 and dopant homogeneity.

A few preliminary LHPG fiber growth experiments by NIGHTINGALE et al. [15] and KWAY et al. [16] have been undertaken to develop techniques for the growth of fiber crystals of Ti^{3+}:Al_2O_3 and to study their properties. In the simplest case, Ti^{3+} doped Al_2O_3 samples cut from bulk crystals produced elsewhere were used as the source material. Due to the small segregation coefficient of Ti^{3+} in Al_2O_3 and perhaps some volatilization of Ti from the molten zone, the concentration of Ti^{3+} in the resulting fiber crystals was very low. A more successful approach involved applying a titanium metal coating to the surface of a pure sapphire source rod by vacuum evaporation. The titanium was then oxidized to TiO_2 and then a fiber grown from this material in a reducing Ar atmosphere. The result was a fiber crystal which did have a measureable Ti^{+3} content, although from preliminary measurements it appeared to be lower than bulk crystals grown by other methods.

Several alternate approaches exist for increasing the dopant concentration. One involves preparing the source rod from hot pressed samples containing a much higher Ti_2O_3 content then found in as-grown crystals. The concentration might be either equivalent to or greater than that put into the melts used for the bulk growth methods.

A more extensive program on the growth and evaluation of Ti^{3+}:Al_2O_3 fiber crystals is underway.

6. Conclusions

The laser-heated pedestal growth method coupled with the growth of small crystals is a very versatile technique. When used for surveying new laser materials it will be faster and more economical than any other crystal growth technology. Laser crystals in fiber form have been successfully grown by this technique from well-known as well as new materials which are difficult to grow in single crystal form, and their properties have been evaluated.

References

1. D. G. Gasson and B. Cockayne, J. Mater. Sci. 5, 100 (1970).
2. J. S. Haggerty, "Production of Fibers by a Floating Zone Fiber Drawing Technique," Final Report NASA-CR-120948, May 1972.
3. C. A. Burrus and J. Stone, Appl. Phys. Lett., 26 318 (1975).

4. J. Stone, C. A. Burrus, A. G. Dentai and B. I. Miller, Appl. Phys. Lett., 29, 37 (1976).

5. J. Stone and C. A. Burrus, J. Appl. Phys., 49, 2281 (1978).

6. C. A. Burrus and L. A. Coldren, Appl. Phys. Lett., 31, 383 (1977).

7. R. S. Feigelson, Proceedings Fifth International Summer School on Crystal Growth, J. Cryst. Growth, to be published.

8. M. Fejer, R. L. Byer, R. Feigelson, and W. Kway, Proceedings of the SPIE, Advances in Infrared Fibers II, 320, 50 (1982).

9. M. Fejer, J. L. Nightingale, and R. L. Byer, U.S. Patent No. 4, 421, 721, Dec. 20, 1983.

10. J. R. Carter and R. S. Feigelson, J. Amer. Ceramic Soc., 47, 141 (1964).

11. W. L. Kway and R. S. Feigelson, unpublished.

12. Kh. S. Bagdasarov, A. A. Kaminsicii, A. M. Kevorkov, and A. M. Prokhorov, Sov. J. Quant. Electron, 4, 927 (1975).

13. M. J. Digonnet, H. J. Shaw, W. L. Kway, and R. S. Feigelson, unpublished.

14. D. L. O'Meara, W. L. Kway, H. J. Shaw, and R. S. Feigelson, unpublished.

15. J. L. Nightingale and R. L. Byer, unpublished.

16. W. L. Kway and R. S. Feigelson, unpublished.

Acknowledgement

This work was initiated and is currently being supported by the NSF-MRL Program through the Center for Materials Research at Stanford University. Part of this work was supported by the Air Force Rome Air Development Center.

Part IV

Vibronic Laser Theory and Cross-Fertilization Through Interdisciplinary Fields

Energy Transfer in Vibronic Laser Materials

R. Orbach

Department of Physics, University of California, Los Angeles, CA 90024, USA

The factors affecting excitation transfer between ions in vibronic laser materials are listed and briefly discussed. Examples are given of the important excitation transfer mechanisms. A new method for utilization of collision-induced electric dipole radiation and collision-induced excitation transfer in the solid state is introduced.

1. Introduction

Excitation transfer for vibronic materials was first formulated by FÖRSTER [1] and DEXTER [2]. Both papers utilized the concept of spectral overlap in their perturbation calculations, with DEXTER [2] introducing multipole interactions coupling the participating ions. It remained for ORBACH [3] and HOLSTEIN et al. [4] to develop the theory from first principles, introducing the phonon coupling and electrostatic coupling on the same footing. Finally, KOHLI and HUANG-LIU [5] included the strong coupling corrections to the microscopic theory.

The purpose of this article is to summarize these treatments, with special reference to vibronic laser materials, and to suggest new methods for excitation transfer utilizing superionic crystalline hosts. The paper is divided into the following Sections: 2. Factors affecting excitation transfer; 3. Examples of excitation transfer processes; 4. Collision-induced matrix elements for radiation and excitation transfers; 5. Summary and acknowledgments.

2. Factors Affecting Excitation Transfer

There are six factors which play a dominant role in excitation transfer between ions in vibronic laser materials. They are:

a. Coupling between ions
b. Coupling of the ions to the lattice
c. Number of lattice quanta which make up the energy mismatch
d. Order of emission/absorption of lattice quanta
e. Site/state for emission/absorption of lattice quanta
f. Grossly different ion-lattice couplings for the ground and excited ionic states.

To establish notation for the discussion of these factors, we define the following quantities:

a. Coupling between ions

We introduce the site-site coupling Hamiltonian between the ions by the symbol H_{S-S}. It can have a multipolar form [e.g., dipole-dipole; dipole-quadrupole; quadrupole-quadrupole] with a rapid fall-off with distance [$1/r^3$; $1/r^4$; $1/r^5$; respectively], or be of an exchange type falling off with distance as $\exp(-\kappa r)$, where κ^{-1} is a characteristic range of interaction. In addition, as detailed by HOLSTEIN et al. [4], excitation transfer can also occur via photon exchange. This process is efficient under conditions

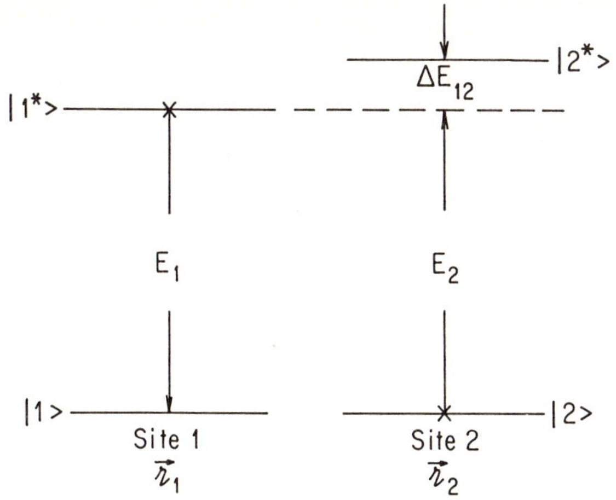

Fig. 1. The excitation transfer process. The ground electronic state at site 1 is denoted by $|1\rangle$; the excited state by $|1^*\rangle$. A similar notation holds for site 2. The symbol × denotes the state of initial occupancy; the arrows the flow of the excitation process. The energy mismatch $\Delta E_{12} = E_2 - E_1$. The sites are located at spatial positions $\vec{r}_1$ and $\vec{r}_2$ with the vector distance denoted by $\vec{r}_{12} = \vec{r}_2 - \vec{r}_1$.

of radiative trapping, with a very long range dependence [falling off as λ/r, where λ is the photon wavelength]. We denote the action of the coupling Hamiltonian by writing its matrix elements as

$$\langle 1,2^*,n_{s,\vec{q}}|H_{s-s}|1^*,2,n_{s,\vec{q}}\rangle = J \qquad (1)$$

where the initial state (before excitation transfer) is specified by site 1 in the excited state [1*], site 2 in the ground state [2], and the vibrational state of the lattice is labeled in terms of the phonon occupation numbers $n_{s,\vec{q}}$ for polarization mode s and wave vector $\vec{q}$. The final state (after excitation transfer) is specified by site 1 in the ground state [1], and site 2 in the excited state [2*]. The coupling Hamiltonian does not itself change the vibrational state of the lattice.

It is clear that the magnitude of J will depend heavily on the character of the initial and final electronic states. If states of the same parity on the same site are involved in the transition, the lowest order matrix element of the electric dipole Hamiltonian will vanish. Electric quadrupole-quadrupole interactions will then dominate the excitation transfer process. If states of different parity are involved, or there is strong opposite parity mixing, the (much stronger) electric dipole-dipole interaction will dominate. Detailed knowledge of the electronic transition in question is essential to determine the strength of the matrix element of the ion-ion coupling defined by Eq. (1).

b. Coupling of the ions to the lattice

The precise origin of the ion-lattice interaction is quite complex. It was VAN VLECK [6] who originally formulated the problem in terms of the normal modes of vibration of the neighboring ligands. Such a symmetry approach assumes that the predominant contributions to the ion-lattice coupling are short range, i.e., only between near neighbors. The local vibrations are then expressed in terms of the running waves (i.e., phonons) of the pure

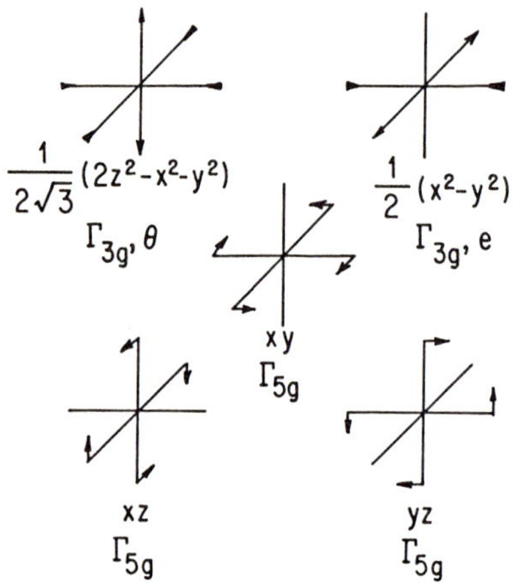

Fig. 2. Normal modes of vibration of octahedral complex of ligand charges. The irreducible representations of even symmetry are denoted by Γ_{ig}, where for the octahedron $i = 3$ and 5. The sub-vectors are exhibited in the usual notation. The arrows denote the direction of ligand motion for each of the sub-vectors.

$$\Gamma_{5g,0} = i(xy)$$

$$\Gamma_{5g,+1} = -\frac{i}{\sqrt{2}}\left[(yz)+i(xz)\right]$$

$$\Gamma_{5g,-1} = +\frac{i}{\sqrt{2}}\left[(yz)-i(xz)\right]$$

crystal. This approach typically assigns parameters for the strength of the dynamic crystalline field coefficients; ab initio calculations being very difficult. The use of group theory can, however, reduce the number of parameters to a minimum [7]. An often used example is an ion surrounded by an octahedral configuration of ligands, pictured in Fig. 2.

The matrix element of ion-lattice coupling Hamiltonian, $H_{ph}(j)$, where the subscript ph means phonon [the understanding is that the local vibrations of Fig. 2 have already been expanded in terms of the running waves], and j denotes the site, is written for the ground state as:

$$\langle j,n_{s,\vec{q}}\pm 1|H_{ph}(j)|j,n_{s,\vec{q}}\rangle = f(j)\ \langle n_{s,\vec{q}}\pm 1||\varepsilon||n_{s,\vec{q}}\rangle \exp(\mp i\vec{q}\cdot\vec{r}_j) \tag{2}$$

where r_j is the position vector of the jth site, and ε represents the relevant component of the strain tensor. The "double bar" matrix element of the strain tensor is given by

$$\langle n_{s,\vec{q}}\pm 1||\varepsilon||n_{s,\vec{q}}\rangle = \sum_{s,\vec{q}} [(\hbar/8MN\omega_{s,\vec{q}})^{1/2}\ (e_{s.\alpha}q_\beta + e_{s,\beta}q_\alpha)] \times \langle n_{s,\vec{q}}\pm 1|a_{s,\vec{q}}^{\dagger} + a_{s,\vec{q}})|n_{s,\vec{q}}\rangle\ , \tag{3}$$

where M, N are the mass of the ion and the total number of lattice sites, respectively, $e_{s,\alpha}$ is the αth component of the polarization vector for the s polarization, and the $a_{s,\vec{q}}^{\dagger}$ ($a_{s,\vec{q}}$) are the creation (destruction) operators for a phonon of polarization index s and wave vector $\vec{q}$. Our treatment will not include crystal anisotropy for simplicity. We shall be concerned only with averages of the strain tensor $\varepsilon_{\alpha\beta}$ over the solid angle Ω. We write:

$$(1/4)\ \langle(e_{s,\alpha}q_\beta + e_{s,\beta}q_\alpha)^2\rangle_\Omega = \alpha_s q^2 \tag{4}$$

where α_s will be of order unity. The remaining matrix element in Eq. (3) generates the usual Bose factors:

$$\langle n_{s,\vec{q}} \pm 1 | a^{\dagger}_{s,\vec{q}} + a_{s,-\vec{q}} | n_{s,\vec{q}} \rangle = \begin{Bmatrix} n(\hbar\omega_{s,\vec{q}}) + 1 \\ n(\hbar\omega_{s,\vec{q}}) \end{Bmatrix}$$

where $n(x)$ is the Bose factor, $n(x) = [\exp(\beta x) - 1]^{-1}$. Above, $\hbar\omega_{s,\vec{q}}$ is the energy of a phonon of polarization index s and wave vector q, and we have made use of the inversion symmetry of the Brillouin zone: $\hbar\omega_{s,\vec{q}} = \hbar\omega_{s,-\vec{q}}$. The symbol β in $n(x)$ stands for $1/k_BT$. We see from Eq. (3) that it is the ion-lattice coupling which changes the vibrational state of the lattice during the excitation transfer process. In an entirely analogous sense, the matrix element of the ion-lattice coupling Hamiltonian is written for the excited state as

$$\langle j^*, n_{s,\vec{q}} \pm | H_{ph}(j) | j^*, n_{s,\vec{q}} \rangle = g(j) \langle n_{s,\vec{q}} \pm 1 || \varepsilon || n_{s,\vec{q}} \rangle \exp(\mp i\vec{q}\cdot\vec{r}_j) \ . \qquad (5)$$

c. Number of lattice quanta which make up the energy mismatch

The energy mismatch between sites 1 and 2 of Fig. 1, ΔE_{12}, must be accounted for by the change in the energy of the vibrational system. We shall consider explicitly one- and two-phonon assisted excitation transfer processes in the next section. For the former, a single phonon makes up for the ionic energy mismatch: $\hbar\omega_{s,\vec{q}} = \Delta E_{12}$. However, as we shall demonstrate in Sec. 3, interference factors act to reduce the magnitude of the single-phonon-assisted excitation transfer process. As a consequence, two-phonon-assisted transfer processes usually dominate. In such cases, it is the difference of phonon energies which makes up the energy mismatch ΔE_{12}. Though such processes are of higher order than single-phonon assisted excitation transfer, reduction of the interference terms makes two-phonon assisted processes dominate the single-phonon assisted process, especially at room temperature.

All of these considerations assume that the energy mismatch ΔE_{12} is within the energy bandwidth of the phonon system. Should ΔE_{12} be significantly greater, then multiple-phonon assistance is necessary. The explicit equations can be derived using the same format as outlined here. One should be warned, however, that multiple-phonon assistance can introduce other factors. See, for example, the article in this volume by BARTRAM [8] for the effectiveness of quadratic coupling when large numbers of phonons participate in the non-radiative relaxation process.

d. Order of emission/absorption of lattice quanta

The excitation transfer process involves both the site-site coupling Hamiltonian and the ion-lattice coupling Hamiltonian (sometimes taken to second order--see 2.c. above). As a consequence, the sequence in which the various perturbations act is very important, and can lead to cancellations in the excitation transfer process. We shall indicate the order of the various steps in the perturbation process explicitly when we come to examine specific excitation transfer processes in Sec. 3. Naturally, this subtlety is not contained in the usual "spectral overlap" calculations for excitation transfer [1, 2].

e. Site/state for emission/absorption of lattice quanta

As can be seen from Eqs. (2) and (5), the strength of the ion-lattice coupling Hamiltonian depends both on the site at which the perturbation takes place [hence, the site label (j)], as well as the state of the ion when the perturbation acts [hence, the different ion-lattice amplitudes f (ground state) and g (excited state) in Eqs. (2) and (5)]. The cancellation mentioned above will be complete (i.e., no excitation transfer); except for the differences between ion-lattice couplings at different site positions, as well as for the differences between ion-lattice couplings in the ground and excited states of the ion at a given site. This will mean, in general, that excitation transfer between different ions (with different ion-lattice couplings) will be more rapid than between identical ions (with the same ion-lattice coupling strengths).

f. Grossly different coupling to the lattice between the ground and excited ionic states.

In addition to the cancellation effect discussed above in 2.3., another factor is important when the ion-lattice coupling is grossly different in the ground and excited state of an ion. This will be a very common situation for vibrationally terminated laser transitions. This is often termed the "Debye-Waller" factor, and was first introduced for excitation transfer by KOHLI and HUANG-LIU [5]. They showed that all of the transition rates calculated for a single-phonon assisted excitation transfer involving weak differences in ion-lattice coupling strengths (e.g., as in Ref. [4]) are to be multiplied by the factor:

$$\exp[-\Phi(T)] \ . \tag{6}$$

In a phonon representation (e.g., an optical phonon of energy $\hbar\omega_s$, where as before s is the polarization index),

$$\Phi(t) = \gamma_s(2n_s + 1) \ , \tag{7}$$

where n_s is the Bose factor appropriate to energy $\hbar\omega_s$, and

$$\gamma_s = [f(j) - g(j)]^2/(\hbar\omega_s)^2 \ . \tag{8}$$

One sees the explicit dependence on the difference in ion-lattice couplings between the ground and excited state in Eq. (8), and on the absolute strength of the ion-lattice coupling in Eq. (6). We plot below the quantity given by Eq. (6) for an optical phonon of energy 300 cm^{-1} at a temperature of 100 K, for various values of the coupling parameter γ_s:

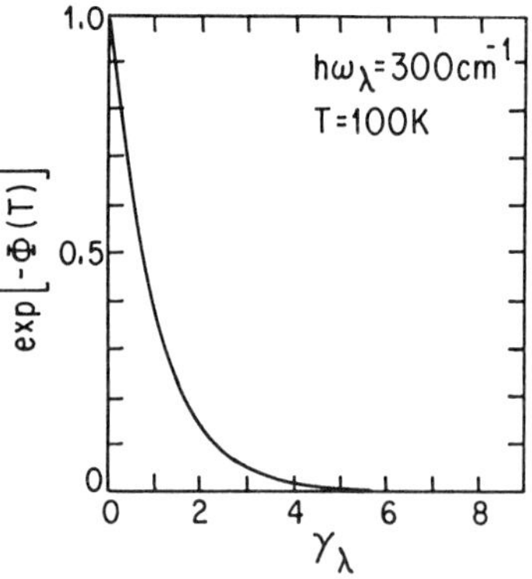

Fig. 3. The "Debye-Waller" reduction factor, Eq. (6), plotted as a function of the coupling parameter γ_λ.

3. Examples of Excitation Transfer Processes

We illustrate in this Section some of the more important excitation transfer processes which involve the perturbations introduced in Section 2. We will not attempt a complete treatment; that can be found in Ref. [4], and in additional references cited therein.

The specific processes we shall exhibit are:

a. Single-phonon assisted transfer
b. Two-site two-phonon assisted non-resonant transfer
c. Single-site two-phonon assisted non-resonant and resonant transfer
d. Single or two-phonon assisted radiative transfer.

We mean by two-site or one-site the spatial position for the ion-lattice interaction which is taken to second order in two-phonon assisted processes. The processes a.-d. appear to span the variety of excitation transfer processes which have been reported in the literature, and which one would expect to be important to vibronic laser transitions. We consider each in turn.

a. Single-phonon assisted transfer

The perturbation sequence is pictured in Fig. 4.

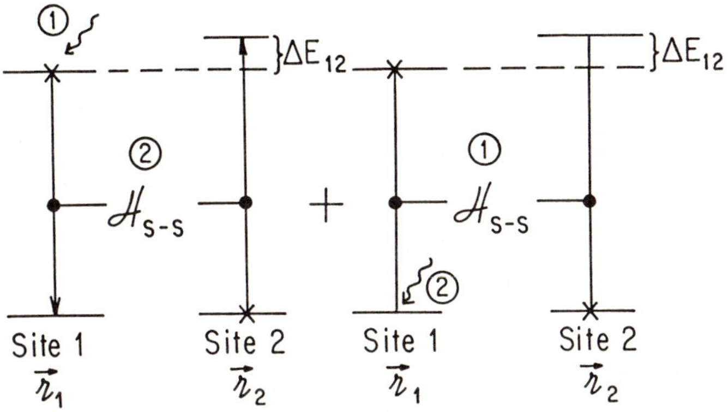

Fig. 4. The perturbation sequence for single-phonon assisted excitation transfer. The circled numbers represent the order of the sequence. Analogous figures for ion-lattice interactions at site 2 have been omitted, but must be added to those exhibited here.

The transition rate for the process pictured in Fig. 4 is given by the "golden rule" expression:

$$W_{2\leftarrow 1} = (2\pi/\hbar) \sum_{s,\vec{q}} |t_{2\leftarrow 1}|^2 \, \delta(\hbar\omega_{s,\vec{q}} - \Delta E_{12}) \,, \tag{9}$$

where $t_{f\leftarrow 2}$ is the "t" matrix, with 2 and 1 representing the final and initial system states, respectively. The delta function at the end of Eq. (9) insures energy conservation. This transition probability per unit time is the excitation transfer rate between ionic sites of differing transition energies (see Fig. 1).

Using all of the expressions for the site-site coupling and ion-lattice Hamiltonians above, and summing over all possibilities (see Fig. 4), we find the "t" matrix which enters Eq. (9) for the single-phonon assisted transfer process pictured in Fig. 4 ($\Delta E > 0$), to equal,

$$t_{2\leftarrow 1} = \frac{J\langle n_{s,\vec{q}}-1|\varepsilon|n_{s,\vec{q}}\rangle \exp(+i\vec{q}\cdot\vec{r}_2)}{-\Delta E_{12}}$$

$$\times \{[f(1)-g(1)] \exp(-i\vec{q}\cdot\vec{r}_{12}) - [f(2)-g(2)]\} \ . \qquad (10)$$

We see that the strength of the single-phonon "t" matrix depends on the difference in ion-lattice coupling constants between the ground and excited levels on each site. Further, for small phonon energies (long phonon wavelengths), the magnitude of the product $\vec{q}\cdot\vec{r}_{12}$ is small ($q = \Delta E_{12}/\hbar v_s$, where v_s is the velocity of sound for polarization s). This means that, if ions 1 and 2 are of a similar type, a near complete cancellation occurs in Eq. (10).

In order to evaluate the full single-phonon assisted transition rate [Eq. (9)], we make the assumption that the ions are identical, apart from their difference in transition energy. For large energy mismatch ($qr_{12} >> 1$):

$$W_{2\leftarrow 1} = \frac{J^2(f-g)^2 \, |\Delta E_{12}|}{\pi\hbar^4\rho} [\textstyle\sum_s (\alpha_s/v_s^5)] \, n(|\Delta E_{12}|)$$

$$= [J^2(f-g)^2/\pi\hbar^4\rho] [\textstyle\sum_s (\alpha_s/v_s^5)] \, k_B T \ , \qquad \beta \, \Delta E_{12} << 1 \ , \qquad (11)$$

where ρ is the crystal mass density. For small energy mismatch ($qr << 1$):

$$W_{2\leftarrow 1} = [J^2(f-g)^2 |\Delta E_{12}|^2 \, r_{12}^2/6\pi\hbar^6\rho] [\textstyle\sum_s (\alpha_s/v_s^7)] \, k_B T \ , \qquad (12)$$

considerably smaller than for large energy mismatch [Eq. (11)]. It is interesting to note that the excitation transfer rate is independent of energy mismatch for large energy mismatch in the high temperature regime. For energy mismatch large compared to k_BT, both large and small energy mismatch conditions lead to a transfer rate which increases with increasing energy mismatch. This is unique to the single-phonon assisted excitation transfer process. We shall see in the next few subsections that the reverse will be true for two-phonon assisted processes.

g. Two-site two-phonon assisted non-resonant transfer

For two-phonon assisted excitation transfer, we have previously noted that only the difference in phonon energies need equal the ionic energy mismatch Thus, for phonons of energy $\hbar\omega_{s',\vec{q}'}$ and $\hbar\omega_{s,\vec{q}}$, energy conservation leads to the condition

$$\hbar\omega_{s',\vec{q}'} - \hbar\omega_{s,\vec{q}} = \Delta E_{12} \ . \qquad (13)$$

These processes are of higher order than one-phonon assisted processes, but they can dominate for two reasons:

(1) limitation of the strength of the single-phonon assisted process because of a small energy mismatch ΔE_{12} (i.e., $\hbar\omega_{s,\vec{q}} = \Delta E_{12}$), and

(2) lack of interference in the two-phonon assisted transfer coherence factor.

The requirement Eq. (13) means that the participating phonons may have energies in the range of k_BT even if $\Delta E_{12} << k_BT$. Such would not be the case for the single-phonon assisted processes. This means that relatively large q phonons are involved in the two-phonon assisted processes so that the interference terms are greatly reduced as compared to the single-phonon assisted transfer process.

The specific process we consider in this subsection is pictured below in Fig. 5.

The details of the calculation are given in Ref. [4]. One finds:

$$W_{2\leftarrow 1} = [J^2(f-g)^4/\pi^3\hbar^7\rho^2] [\textstyle\sum_s (\alpha_s/v_s^5)]^2 \, (k_BT)^3 \ . \qquad (14)$$

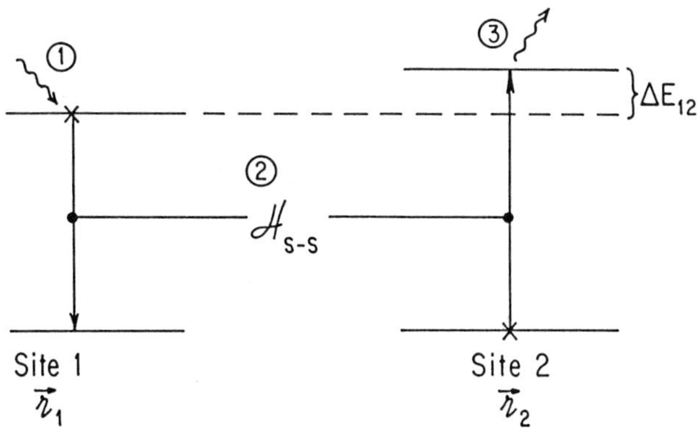

Fig. 5. The two-site two-phonon non-resonant assisted transfer process. The circled numbers represent the order of the perturbation sequence. There are a total of sixteen processes of this kind (i.e., different order, and action of the ion-lattice interaction at different ionic levels).

As promised, the excitation transfer rate is two orders higher in the ion-lattice interaction. We have assumed that the temperature is low compared to the Debye temperature, and we have ignored interference terms. These have been included explicitly by HAMILTON et al. [9]. Where higher lying states are not close by, this process has been seen to dominate the excitation transfer process in physical systems [9]. Note from Fig. 5 that the phonon absorption and emission takes place at different sites. As a consequence, the transfer process cannot be formulated in terms of a spectral overlap. Thus, this process cannot be derived using the methods of FORSTER [1] and DEXTER [2]. One must treat all the perturbation terms on the same level. Note that the transfer rate, Eq. (14), is independent of the energy mismatch ΔE_{12}.

c. Single site two-phonon assisted non-resonant and resonant transfer

These two separate processes are pictured below in Fig. 6. Again, the de-

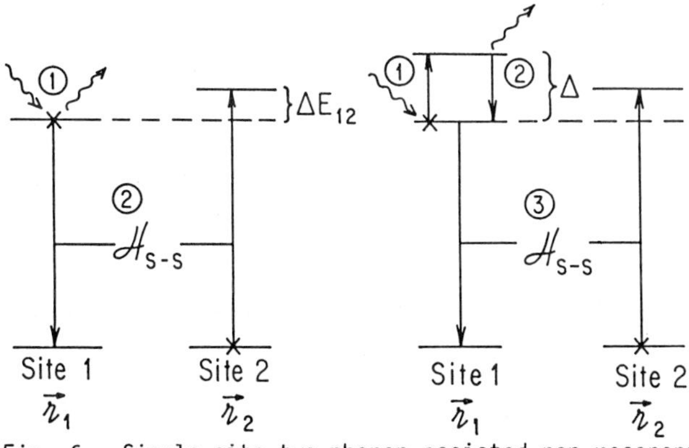

Fig. 6. Single site two-phonon assisted non-resonant and resonant transfer. The circled numbers represent the order of the perturbation sequence. There are a total of four processes of the one site Raman process type (left-hand figure), and six of the one site resonance process type (right-hand figure).

tails of the calculation can be found in Ref. [4]. The similarities between the two processes exhibited above are that the transition rate is proportional to $J^2/(\Delta E_{12})^2$. This demonstrates that the higher order single site two-phonon assisted excitation transfer rates are inversely proportional to the square of the energy mismatch. Hence, their importance falls off as ΔE_{12} increases in magnitude, relative to the two-site two-phonon assisted non-resonant transfer process which has been shown to be independent of ΔE_{12} [see Eq. (14)].

The differences between the two processes are to be found in the temperature dependences. The single site two-phonon assisted non-resonant transfer process is proportional to T^7, the resonance process to $\exp(-\Delta/k_BT)$ where Δ is the splitting within the excited states. Because of the single-site character of both of these processes, the action of the ion-lattice Hamiltonian is the same that one would find in a calculation of the optical transition linewidth. Hence, the excitation transfer rate can be formulated in terms of a spectral overlap, and as such would be implicitly contained in a FÖRSTER and DEXTER formulation of the excitation transfer rate.

d. Single or two-phonon assisted radiative transfer

This peculiar excitation transfer process is most effective when the photons released with the ionic transition energies E_1 or E_2 (see Fig. 1 for definitions) are trapped. Radiative trapping conditions lead to a highly efficient excitation transfer; every photon emitted participates in the excitation transfer process. The one-phonon contribution is pictured in Fig. 7. The transition rate calculated for phonon-assisted radiative transfer [4] is proportional to $[(1/\tau_{rad})(f-g)^2/(\Delta E_{12})^2]n(|\Delta E_{12}|)$, where τ_{rad} is the radiative lifetime. This is because the effective site-site coupling constant for the strong trapping regime, as has been stated

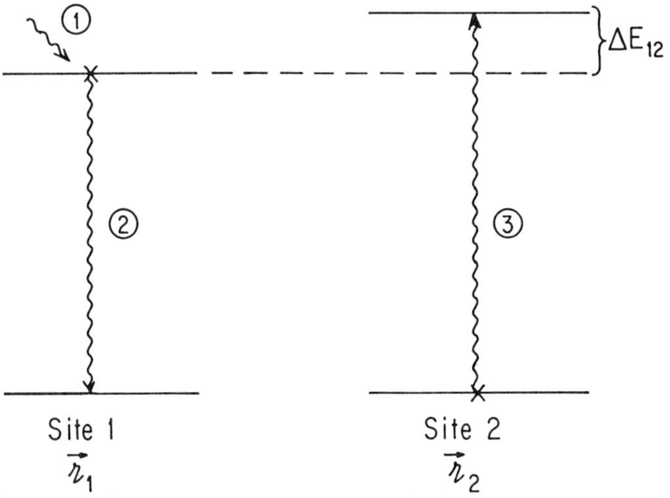

Fig. 7. Single-phonon assisted radiative transfer. The circled numbers represent the order of the perturbation sequence. A phonon of wave vector $\vec{q}$ is absorbed at site 1, a photon of wave vector $|\vec{k}| = E_2/\hbar c$ is emitted at site 1 (question: how did the photon "know" it had an energy equal to the transition energy at site 2?), and the same photon is absorbed at site 2. An analogous figure contributes for phonon absorption in the ground state at site 1, and two more for phonon absorption at site 2.

previously, equals $(\hbar/\tau_{rad})/(r_{12}/\lambda)$, where τ_{rad} is the radiative lifetime and λ is the photon wavelength. This result exhibits the very high efficiency of this process: the range dependence of the site-site coupling is very large, falling off only as $1/r_{12}$. As noted in the figure caption, the physics is anything but classical. In order: a phonon of arbitrary energy is absorbed, a photon of arbitrary energy is emitted, and a photon of specified energy is absorbed. The last act fixes the energy of the emitted photon, and forces the phonon energy to be equal to the energy mismatch.

This completes our treatment of excitation transfer processes for fixed ionic sites. Space limitations have made our treatment necessarily brief. Further elaboration of many of the ideas discussed here can be found in Ref. [4].

Our treatment has so far relied on multipolar, exchange, or radiative coupling between fixed ionic sites. Very recent developments [10] using solid state host materials which are super-ionic conductors allow for dynamic ionic motion in the excitation transfer process. We discuss this interesting possibility below.

4. Collision-Induced Matrix Elements for Radiation and Excitation Transfer

A very recent paper by JANSEN et al. [10] reported laser action of Nd^{3+} ion exchanged into a platelet of Na^+ beta" alumina: $Na_{1+x}Mg_xAl_{11-x}O_{17}$. The remarkable properties of this material, beyond the fact of lasing action itself, lie in the ability to dope various regions of the same preferentially with active ions because of the ion transport behavior of the material [10]. Another feature has to do with the possibility of ion-exchanging another ion, say, Cr^{3+}, into the material in addition to the Nd^{3+}. Hence, one can conceive of optically pumped mobile Cr^{3+} ions transferring their excitation by ionic collision with the fixed Nd^{3+} sites. This should be a much stronger mechanism than that arising from electric multipolar interactions between ions. The transfer scheme is pictured below in Fig. 8.

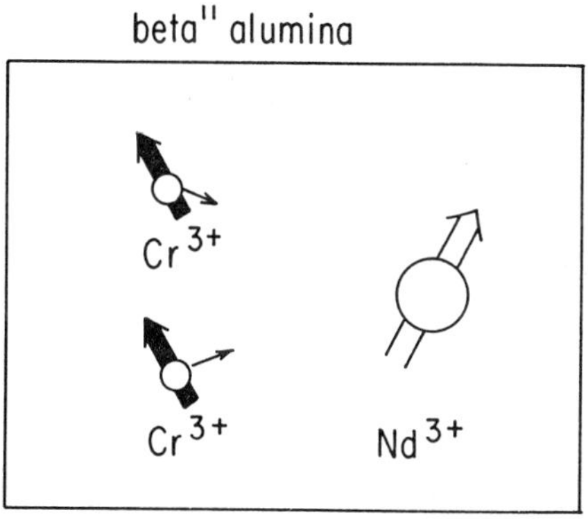

Fig. 8. A pictoral description of collision-induced excitation transfer between the (mobile) pumped Cr^{3+} and the (fixed) Nd^{3+} in beta" alumina.

Another feature which could also play a role is collision-induced matrix elements for the optical emission transition itself. This is a well known phenomenon in gases. In solids with mobile ions there is no reason why similar processes could not occur. For example, POWELL [11] reported that Rh^{2+} ion exchanged into beta" alumina possessed too small an optical transition matrix element for effective laser action. One can speculate on introducing a charged mobile ion by ion-exchange which, upon collision with the optically pumped Rh^{2+}, would induce an oscillator strength which would allow the fixed Rh^{2+} to lase. Presumably, an external electric field applied to the material could affect the diffusion rate of the mobile ion (biased hopping). One could then obtain a time dependent electric field modulation of the Rh^{2+} fluorescent lifetime or laser oscillator strength.

The mobility of ions in the beta" alumina thus allows one to use collision-induced excitation transfer and induced dipole matrix elements which have been so highly developed for gaseous systems, but in a solid state material where the densities and stabilities are much higher.

5. Summary and Acknowledgments

We have tried to summarize in this article the major aspects of excitation transfer processes relevant to vibronic laser materials. We have, in Section 3., outlined the various excitation transfer processes which we expect play an important role in vibronic laser materials. The important parameters have been introduced in Section 2. They can be incorporated directly into our final expressions in Section 3 as appropriate for the specific physical system. Thus, when strong ion-lattice coupling differences exist between ground and excited states, as expected for many vibronic laser transitions, the Debye-Waller reduction of the transfer rate can be obtained by simply multiplying the rate calculated in Section 3. by Eq. (6). In the space available here, not much more than an outline of ideas could be presented. However, it is hoped that the physics has been made sufficiently clear that the features of excitation transfer relevant to vibronic laser design are clear. In addition to describing known excitation transfer processes, some new thoughts are presented which may allow the techniques pioneered in gases for collision-induced electric dipole moments and excitation transfer to be used in the solid state.

The author is indebted to Professor B. Dunn for his helpful insights into the beta" alumina system. This work was supported by the National Science Foundation, grant No. DMR 81-15542.

References

1. T. Förster: Ann. Phys. (Paris) 2, 55 (1948)
2. D. L. Dexter: J. Chem. Phys. 21, 836 (1953)
3. R. Orbach: In Optical Properties of Ions in Crystals, ed. by H. M. Crosswhite, H. W. Moos (Interscience, New York 1967) p. 445
4. T. Holstein, S. K. Lyo, R. Orbach: In Topics in Applied Physics, Vol. 49: Laser Spectroscopy of Solids, ed. by W. M. Yen, P. M. Selzer (Springer-Verlag, Berlin-Heidelberg-New York 1981) p. 39
5. M. Kohli, N. L. Huang-Liu: Phys. Rev. B 9, 1008 (1974)
6. J. H. Van Vleck: Phys. Rev. 59, 724 (1941)
7. R. Orbach and H. Stapleton: In Electron Paramagnetic Resonance, ed. by S. Geschwind (Plenum Press, New York-London 1972) p. 121
8. R. Bartram: In 1st Annual Conference on Tunable Solid State Lasers, ed. by P. Hammerling (Springer-Verlag, Berlin-Heidelberg-New York 1984)
9. D. S. Hamilton, P. M. Selzer, W. M. Yen: Phys. Rev. B 16, 1858 (1977); D. S. Hamilton: Ph.D. Thesis, University of Wisconsin, Madison (1977) unpublished
10. M. Jansen, A. Alfrey, O. M. Stafsudd, B. Dunn, D. L. Yang, G. C. Farrington: Optics Lett. 9, 119 (1984)
11. R. Powell: Comments presented at the meeting, "1st Annual Conference on Tunable Solid State Lasers," 13-15 June, 1984, LaJolla Institute, LaJolla, California.

Theory of Fluorescence Quenching in Low-Field Chromium Complexes in Solids

Ralph H. Bartram

Department of Physics and Institute of Materials Science
University of Connecticut, Storrs, CT 06268, USA

Low crystal-field octahedral complexes of Cr^{3+} exhibit broadband ($^4T_{2g} \rightarrow {}^4A_{2g}$) fluorescence, but their potential application to tunable lasers is limited by host-dependent thermal quenching of fluorescence. Several theoretical models for radiationless transitions in $Cs_2NaYCl_6:Cr^{3+}$, whose parameters are constrained by experimental data on the temperature dependence of fluorescence spectra and lifetimes, are compared. Single configuration-coordinate and multi-mode linear coupling models fail by many orders of magnitude. A new model, which combines linear coupling to an a_{1g} mode with quadratic coupling to a t_{2g} mode within the harmonic approximation, is much more successful.

1. Introduction

Low-field chromium complexes in solids provide potential materials for tunable, solid-state infrared lasers, but their application is limited by thermal quenching of fluorescence, which is found to be strongly host-dependent [1]. The object of the present investigation is to achieve an understanding of the factors which govern thermal quenching. Host crystals employed in the investigation are ordered perovskites (elpasolites), which can accommodate a substitutional trivalent cation impurity at a site of rigorous octahedral symmetry, without charge compensation. The present discussion will concentrate on one of these, $Cs_2NaYCl_6:Cr^{3+}$. Several theoretical models, with parameters constrained by experimental data on the temperature-dependence of fluorescence spectra and lifetimes, are compared.

2. Fluorescence Spectra and Lifetimes

The ground state of Cr^{3+} is $^4A_{2g}$. The two lowest excited states are ^2E_g, which belongs to the same strong-field configuration (t_2^3) as the ground state, from which it is split by exchange interaction, and $^4T_{2g}$ (t_2^2e) which is split from the ground state by the crystal field. High- and low-field complexes are distinguished by their luminescence behavior: High-field complexes, such as ruby, in which the ^2E_g state lies lower, exhibit narrow-band phosphorescence (spin-forbidden). Low-field complexes, in which the $^4T_{2g}$ state lies lower, exhibit broad-band fluorescence (spin-allowed), which is of potential utility in tunable lasers.

Laser-induced fluorescence spectra and lifetimes were measured as functions of temperature on polycrystalline samples of low- and intermediate-field compounds $Cs_2NaYCl_6:Cr^{3+}$, $K_2NaScF_6:Cr^{3+}$ and $K_2NaGaF_6:Cr^{3+}$. The lifetime data are plotted in Fig. 1 on a semi-log scale. Two distinct temperature ranges are apparent for each compound: The slowly declining lifetime in the low-temperature range reflects a combination of phonon assistance of parity-forbidden transitions and competition between ^2E_g and $^4T_{2g}$ luminescence; while the much more rapid decline in the high-temperature range, which correlates with diminishing quantum efficiency, reflects the dominance of competing non-radiative transitions. Sensitivity of ther-

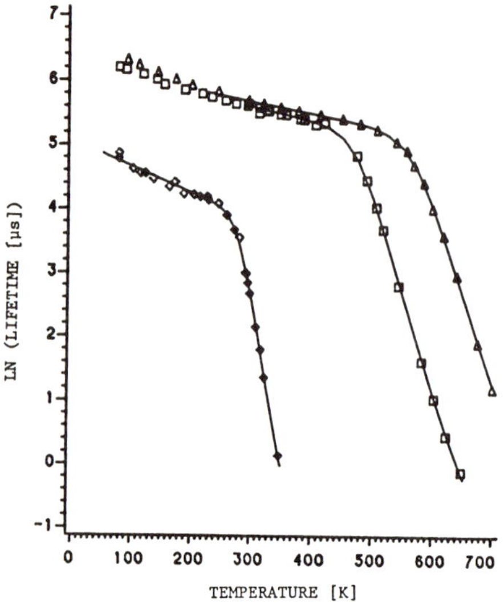

Fig. 1. Fluorescence lifetimes as a function of temperature for $Cs_2NaYCl_6:Cr^{3+}$ (diamonds), $K_2NaScF_6:Cr^{3+}$ (squares) and $K_2NaGaF_6:Cr^{3+}$ (triangles)

mal quenching to host crystals, and even to constituents beyond the immediate ligands, is evident from these data; similar sensitivity has been reported for V^{2+} in fluoride compounds as well [2].

The lifetime data of Fig. 1 were fitted over a limited temperature range to an expression of the form

$$\ln(\tau) = -\ln(\tau_R^{-1} + \tau_{NR}^{-1}), \tag{1}$$

where the radiative rate τ_R^{-1} was assumed, for convenience of extrapolation, to have temperature dependence

$$\tau_R^{-1} = \tau_0^{-1} \exp(\alpha T), \tag{2}$$

and the non-radiative rate was taken to be

$$\tau_{NR}^{-1} = s \exp(-\Delta E/k_B T). \tag{3}$$

Frequency factors s and activation energies ΔE are listed in Table 1.

Table 1. Frequency factors s and activation energies ΔE for radiationless transitions, inferred from fit of lifetime data to (1)-(3)

Compound	$s(10^{13} sec^{-1})$	$\Delta E(cm^{-1})$
$Cs_2NaYCl_6:Cr^{3+}$	3.8	4250
$K_2NaScF_6:Cr^{3+}$	1.2	7270
$K_2NaGaF_6:Cr^{3+}$	5.2	9240

Broad-band fluorescence spectra provide empirical information which constrains theoretical models for radiationless processes. In particular, parameters which characterize adiabatic potential energy surfaces are derived from the temperature-dependence of first and second moments of the fluorescence spectra.

3. Promoting Interaction

In the case of internal conversion, non-radiative transitions between Born-Oppenheimer states are mediated by the non-adiabaticity operator [3]. First-order perturbation theory leads to an expression for the non-radiative transition rate W_{NR} of the form

$$W_{NR} = \nu\omega[\bar{n}G(\Omega_0+\omega)+(\bar{n}+1)G(\Omega_0-\omega)]. \qquad (4)$$

It is convenient to distinguish between promoting modes, which mix the initial and final electronic states, from accepting modes, which absorb their energy difference. Coupling to accepting modes is reflected in the normalized lineshape function $G(\Omega)$, where Ω is measured from the zero-phonon transition. The promoting-mode interaction is incorporated in the factor $\nu\omega$, where ω is the promoting-mode frequency. The phonon occupation number $\bar{n}$ is given by

$$\bar{n} = [\exp(\hbar\omega/k_BT)-1]^{-1}. \qquad (5)$$

A promoting mode of t_{1g} symmetry is required to mix ${}^4T_{2g}$ and ${}^4A_{2g}$ electronic states. Counter-rotating displacements of anion octahedra about trivalent and monovalent cations provide a strongly coupled mode of the required symmetry [4], contrary to earlier speculation concerning promoting-mode selection rules [5]. For want of more precise information, we have assumed that the frequency of the t_{2g} promoting mode is the same as that of the a_{1g} accepting mode; i.e. $\omega = \omega_{t1g} = \omega_{a1g}$. The promoting-mode factor ν was calculated from a point-ion model for the ligands in the non-Condon approximation advocated by Huang [6],

$$\nu = (2\pi^2/M\hbar\omega_{a1g}{}^2)\,|\langle\zeta|V_z(t_{1g})|v\rangle|^2, \qquad (6)$$

$$V_z(t_{1g}) = -(35/2r_o{}^6)xy(x^2-y^2), \qquad (7)$$

where M is the anion mass and r_o the chromium-ligand distance. Matrix elements were evaluated with free-ion d orbitals of gaussian form [7]. The calculated value for $Cs_2NaYCl_6:Cr^{3+}$, $\nu = 1.6 \times 10^{12}\,s^{-1}$, is about what one would have guessed; Strück and Fonger have assumed the value $\nu \cong 10^{14}\,s^{-1}$ in a related calculation for ruby and emerald [8].

4. Linear-Coupling Models

The central problem in the application of radiationless-transition theory is evaluation of the normalized lineshape function $G(\Omega)$ in (4) for values of its argument, $\Omega = \Omega_o \pm \omega$, which are very far from the range of arguments accessible to direct observation. The exact form of $G(\Omega)$ depends on the model assumed for accepting mode interactions, but the parameters of each model are severely constrained by data on the temperature-dependence of fluorescence spectra.

The simplest theoretical model for radiationless de-activation of low-field chromium complexes is based on linear coupling to a single accepting-mode configuration coordinate, which we can identify with a symmetrical (a_{1g}) displacement of the ligands. This model is illustrated in Fig. 2, which shows the adiabatic potential energy curves for the ${}^4A_{2g}$ and

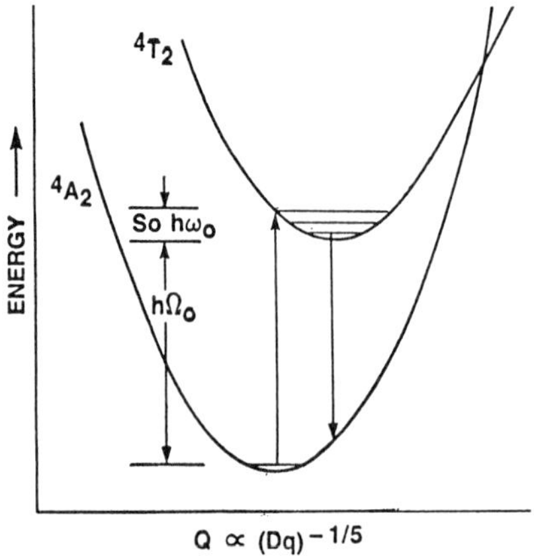

Fig. 2. Configuration coordinate diagram for single-coordinate, linear coupling model

$^4T_{2g}$ states as functions of the configuration coordinate Q. The parameters of this model are the energy gap $E_o = \hbar\Omega_o$, the Huang-Rhys factor S_o (the mean number of phonons excited in a Frank-Condon transition from the ground state), and the a_{1g} phonon energy $\hbar\omega_o$, which is the same in both electronic states. These parameters are completely determined by the temperature-dependence of the first and second moments of fluorescence spectra.

The lineshape function $G(\Omega)$ was calculated from the formula of Huang and Rhys [9],

$$G(\Omega)=\exp[-S_o(2\bar{n}+1)]$$

$$\times \sum_{m=-\infty}^{\infty}[(\bar{n}+1)/\bar{n}]^{m/2} I_{|m|}(2S_o[\bar{n}(\bar{n}+1)]^{1/2})\,\delta(\Omega-m\omega_o), \quad (8)$$

in a smoothed approximation. The Huang-Rhys theory has been very successful in elucidating trends in limiting cases. It predicts the energy-gap law [10] in the weak-coupling limit (S_o<<1), which has been amply verified for rare-earth impurities, and it reduces to the Mott formula [11] in the high temperature, strong-coupling limit ($T \gg \hbar\omega_o/k$, $S_o \gg 1$), appropriate to F centers. The complete formula (8) is required in the present case, which is an example of intermediate coupling.

With the calculated value of ν, the single-mode, linear-coupling model leads to a predicted non-radiative transition rate for $Cs_2NaYCl_6:Cr^{3+}$ which is smaller than the observed rate by a factor of 10^{13} at 300K. Furthermore, the predicted activation energy is much too high as well (by about a factor of two). The single- configuration coordinate model of Struck and Fonger for ruby and emerald [8] fails in the same way when the phonon frequency is constrained by optical spectra.

In an investigation of Co^{2+} in $KMgF_3$, Sturge [12] extended the linear coupling model to include a range of phonon frequencies, following the theoretical formulation of Pryce [13]. This extension was motivated by the expectation that the higher frequency phonons would make a disproportionately large contribution to the non-radiative transition rate. Sturge's calculation was fairly successful, but a residual discrepancy was attributed to anharmonicity, whose efficacy was demonstrated by a hypothetical example.

We have considered a similar multimode, linear-coupling model as an alternative theoretical model for low-field chromium complexes in elpasolites. In this calculation, we have employed the approximate lineshape function of Weissman and Jortner [14],

$$G(\Omega)=\exp[-S_o(2\bar{n}+1)]$$

$$\times \sum_{m=-\infty}^{\infty} [(\bar{n}+1)/\bar{n}]^{m/2} I_{|m|}(2S_o[\bar{n}(\bar{n}+1)]^{1/2}) F(m+k_o,k_o,\Omega), \quad (9)$$

$$F(m+k_o,k_o,\Omega) = [2\pi(m+2k_o)\sigma_o^2]^{-1/2}\exp[-(\Omega-m\omega_o)^2/2(m+2k_o)\sigma_o^2], \quad (10)$$

$$k_o = (1/2)\{-m + [m^2 + 4S_o^2\bar{n}(\bar{n}+1)]^{1/2}\} \quad (11)$$

$$\omega_o = S_o^{-1}\int_0^{\infty} d\omega\,\omega A(\omega), \quad (12)$$

$$\sigma_o^2 = S_o^{-1}\int_0^{\infty} d\omega\, \omega^2 A(\omega)-\omega_o^2, \quad (13)$$

where $A(\omega)$ is a single phonon sideband. The Weissman-Jortner theory is a refinement of the Pryce theory which emphasizes the net number m of phonons emitted in the transition, and which provides an improved approximation at finite temperatures. This multimode extension of linear coupling theory introduces an additional parameter, associated with the range of phonon frequencies, which we have taken to be

$$\delta \cong \sigma_o^2/\omega_o^2; \quad (14)$$

i.e., the ratio of the second moment of the single-phonon sideband to the square of its first moment. The adjusted value of this parameter which yields the correct activation energy, $\delta = 0.3$, is physically reasonable. However, the absolute non-radiative transition rate predicted by this multi-mode, linear-coupling model is still too small by a factor of 10^{10}. Although it is conceivable that a discrepancy of this size could also be attributed to anharmonicity (e.g., Sturge's example [12]), we believe that it is imperative to exhaust the alternative possibilities within the harmonic approximation.

5. Quadratic-Coupling Model

The failure of linear-coupling models suggests consideration of quadratic coupling; i.e. different vibration frequencies in the two electronic states. Although quadratic coupling to a_{1g} modes is possible in principle, there is little reason to expect it to be significant, either from the Tanabe-Sugano diagram [15] or from the observed symmetry of absorption and emission bands. However, there is ample spectral evidence for coupling to modes of lower symmetry; in particular, e_g and t_{2g} modes. Since these lower symmetry displacements remove the orbital degeneracy of the $^4T_{2g}$ state, a complete description of just the $^4A_{2g}$ and $^4T_{2g}$ states involves four adiabatic potential energy surfaces in a six-dimensional configuration-coordinate space.

For simplicity, we have adopted a model which retains just one orbital component of the $^4T_{2g}$ state in addition to the $^4A_{2g}$ state, and just one of the three t_{2g} displacements together with the a_{1g} displacement; i.e. two adiabatic potential energy surfaces in a two-dimensional configuration-coordinate space. The selected t_{2g} displacement is that which is maximally effective in diminishing the crystal-field splitting, as illustrated in Fig. 3. Only those d orbitals which differ in the $^4A_2$ and $^4T_2$ states are

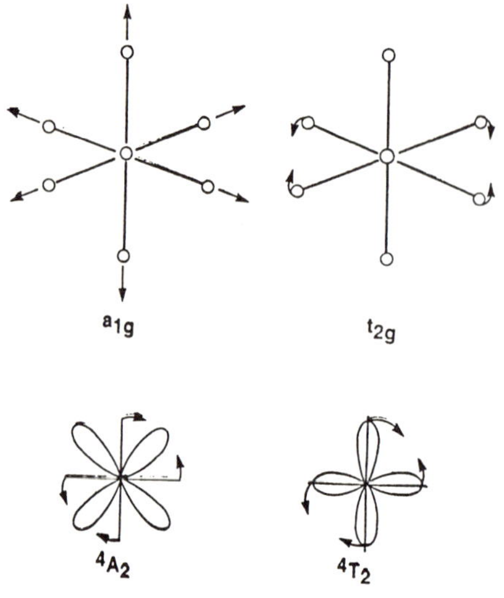

Fig. 3. Two configuration-coordinate model, with linear coupling to an a_{1g} mode and quadratic coupling to a t_{2g} mode

shown; the lobes of the $t_{2g}(\zeta)$ orbital in the $^4A_2$ state avoid the ligands, whereas those of the $e_g(v)$ orbital in the $^4T_2$ state are directed toward the ligands. This difference is the source of the crystal field splitting, and it is evident from Fig. 3 that a t_{2g} displacement enhances the energy of the ζ orbital and diminishes that of the v orbital, independent of its sign, with consequent quadratic coupling. Thus our model incorporates pure quadratic coupling to one t_{2g} coordinate, and pure linear coupling to one a_{1g} coordinate. This two-coordinate model is clearly oversimplified, and, in particular, it suppresses the Jahn-Teller effect. However, we believe that it retains the features essential to an explanation of the observed high non-radiative transition rate.

Adiabatic potential energy curves, calculated from a point-ion crystal field model, are given in the harmonic approximation by

$$E(^4A_{2g}) = -12Dq + (1/2)M\omega_{t2g}^{2}Q_{t2g}^{2}, \tag{15}$$

$$E(^4T_{2g}) = -2Dq + (1/2)M\omega'_{t2g}{}^{2}Q_{t2g}^{2}, \tag{16}$$

$$\omega'_{t2g}{}^{2} = \omega_{t2g}^{2} - 105Dq/2Mr_o^{2}. \tag{17}$$

It is evident that the only new parameter in this model is the t_{2g}-mode phonon energy $\hbar\omega_{t2g}$ in the ground electronic state, since the strength of quadratic coupling is determined by (17). Normalized lineshape functions were calculated for each mode by the method of Struck and Fonger [16], in which individual vibrational overlap integrals are evaluated. The composite lineshape was then calculated by numerical convolution of the linear (a_{1g}) and quadratic (t_{2g}) components,

$$G(\Omega) = \int G_{a1g}(\Omega-\Omega')G_{t2g}(\Omega')d\Omega'. \tag{18}$$

The parameters of the model were adjusted to fit simultaneously the temperature-dependence of the second moment of the fluorescence lineshape and the activation energy of the radiationless transition rate. Optimum values of the parameters for $Cs_2NaYCl_6:Cr^{3+}$ are listed in Table 2, where the phonon energies are seen to compare favorably with values inferred from low-temperature spectra for a related compound, $Cs_2NaInCl_6:Cr^{3+}$. The optimum value of the promoting interaction parameter ^{2} is also listed in Table 2, and is seen to be larger than the calculated value by a factor of 10^3. Thus the present two-mode model with quadratic coupling is better than the single-mode linear-coupling model by a factor of 10^{10}, and better than the multimode, linear- coupling model by a factor of 10^7.

Table 2. Optimum parameters of the two-coordinate, quadratic coupling model for $Cs_2NaYCl_6:Cr^{3+}$

parameter	adjusted value	experiment ($Cs_2NaYCl_6:Cr^{3+}$)[a]
S_o	6.03	-
ν^o	1.5×10^{15}	-
ω_{a1g}	237 cm^{-1}	294 cm^{-1}
ω_{t2g}	119 cm^{-1}	139 cm^{-1}
ω_{t2g}	(70 cm^{-1})[b]	-

a. Gudel and Snellgrove, Inorg. Chem. 17, 1617 (1978)
b. Calculated ab initio from other parameters

6. Discussion

It is evident that the quadratic coupling model is far more successful than the single-coordinate, linear-coupling model. In order to explain this advantage, it is necessary to examine the lineshape simulations more closely.

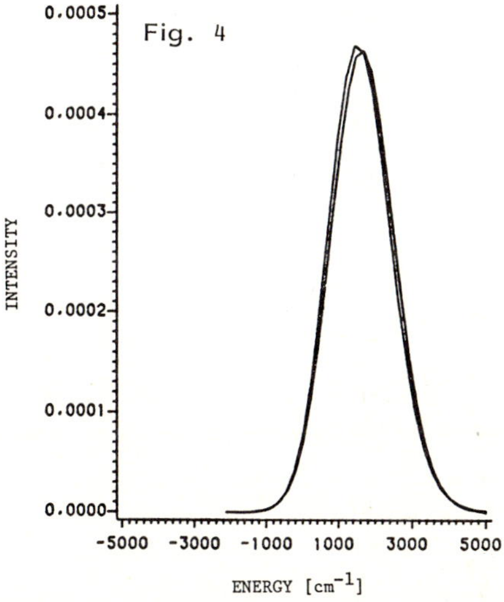

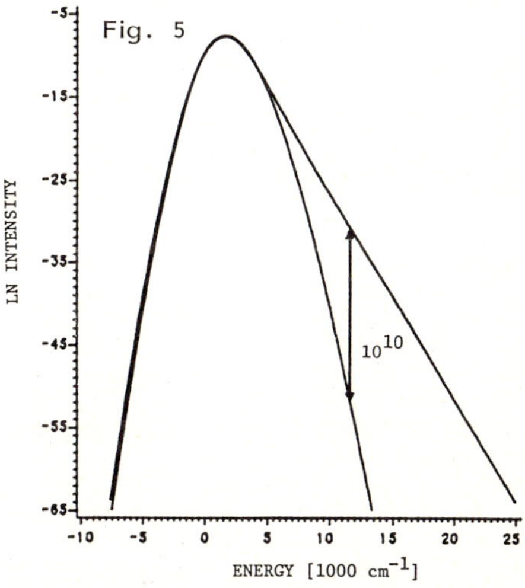

Fig. 4. Comparison of simulated lineshape functions for single-coordinate, linear-coupling model and two-coordinate, quadratic coupling model.
Fig. 5. Comparison of the same lineshape functions as in Fig. 4, on a semi-log scale

Lineshape simulations are compared in Fig. 4 for the single configuration- coordinate, linear-coupling model and the two-coordinate model with quadratic coupling. The two lineshapes appear to be virtually indistinguishable over that part of the energy range which is accessible to direct observation, and either one is in satisfactory agreement with the measured fluorescence spectrum.

Comparison of the same two lineshape functions on a semilog scale over a much wider energy range, as shown in Fig. 5, is more revealing, however. At the gap energy, $\hbar\Omega_0$, the two lineshape functions differ by a factor of 10^{10}, which accounts for the difference in predicted radiationless transition rates. The origin of this difference can be understood from a classical model [16], in which the lineshape function for linear coupling is approximately gaussian, while that for quadratic coupling is approximately exponential. The quantum mechanical lineshapes are similar to the classical lineshapes, and are thus approximately parabolic and linear, respectively, on a semilog scale. The linear tail characteristic of quadratic coupling is preserved in the convolution of lineshape functions in the two-coordinate model, with consequent relative enhancement of the composite lineshape function at high energy.

The two-coordinate, quadratic coupling model is also far more successful than the multi-mode, linear coupling model (by a factor of 10^7), even though both models contain the same number of adjustable parameters. The residual discrepancy of 10^3 presumably reflects the inadequacy of our simplifications, including the restriction to two modes and two electronic states, as well as the assumed harmonic approximation.

7. Acknowledgements

This investigation was supported in part by an NSF Industrial Research Participation Grant, and in part by U.S. Army Research Office Contract No. DAAG29-82-K-0158. The author is pleased to acknowledge the collaboration of Dr. Alexander Lempicki, Dr. Leonard Andrews and Dr. B.C. McCollum of GTE Laboratories, Inc., in the experimental aspects of this research, and of Mr. John Charpie of the University of Connecticut in the computational aspects. Thanks are due to Mr. Carmen Giunta, Dr. Peter Kenyon and Dr. Sylvie Desjardins for experimental assistance, and to Dr. William Miniscalco for helpful discussions.

References

1. P.T. Kenyon, L. Andrews, B. McCollum and A. Lempicki: IEEE J. Quantum Electronics QE-18, 1189 (1982).
2. P.F. Moulton and R.E. Fahey: Solid State Research Report, Lincoln Laboratory, M.I.T. (1983:1), pp. 21-23.
3. R. Englman: Non-Radiative Decay of Ions and Molecules in Solids, North-Holland (1979).
4. A. Lentz: J. Phys. Chem. Solids 35, 827 (1974).
5. L.J. Andrews, A. Lempicki and B.C. McCollum: Chem. Phys. Lett. 74, 404 (1980).
6. K. Huang: Scientia Sinica 24, 27 (1981).
7. F.J. Hay: J. Chem. Phys. 66, 4377 (1977).
8. W.H. Fonger and C.W. Struck: Phys. Rev. B11, 325 1 (1975).
9. K. Huang and A. Rhys: Proc. Roy. Soc. (London) A204, 406 (1950).
10. F. Auzel: in Luminescence of Inorganic Solids, ed. B. DiBartolo (Plenum, NY 1977) p. 67.

11. N.F. Mott: Proc. Roy. Soc. (London) A167, 384 (1938).
12. M.D. Sturge: Phys. Rev. B8, 6 (1973).
13. M.H.L. Pryce: in Phonons, ed. R.W.H. Stevenson (Plenum, N.Y. 1966).
14. Y. Weissman and J. Jortner: Phil. Mag. B37, 24 (1978).
15. S. Sugano, Y. Tanabe and H. Kamimura: Multiplets of Transition-Metal Ions in Crystals (Academic Press, 1970) p. 109.
16. C.W. Struck and W.H. Fonger: J. Luminescence 10, 1 (1975).
17. F.E. Williams, J. Chem. Phys. 19, 457 (1951).

Theoretical Methods for the Study of Transition Metals in Crystals

N.W. Winter

Lawrence Livermore National Laboratory, Livermore, CA 94550, USA

R.M. Pitzer

Department of Chemistry, Ohio State University, Columbus, OH 43210, USA

I. Introduction

The characterization of the electronic structure of impurities in solids impacts a large class of phenomena and devices. It is particularly important for the systematic development of new solid state laser media. The large number of ion-host combinations that must be studied in order to optimize the linear and non-linear optical properties and the thermal and mechanical properties of the laser preclude a case-by-case experimental investigation. This suggests that an aggressive theoretical effort to determine the optical properties of these materials will be useful in the search for new tunable solid state lasers.

Unlike the Bloch states of a perfect crystal, the electronic states associated with impurities are localized on a very few atoms. Because their energy levels are usually well separated from those of the host crystal, the electronic spectroscopy of transition metal and rare earth ion impurities offers a wealth of information. The theoretical methods used to describe the properties of transition metals in crystals do not differ in principle from those used to treat atoms and molecules. The ion impurity and the neighboring crystal atoms define a molecular cluster that can be studied with a variety of empirical, semi-empirical, and ab initio methods. These include crystal field theory, semi-empirical molecular orbital theory, X-alpha and ab initio methods. The purpose of our study is to accurately locate the position of excited electronic states, calculate transition probabilities for absorption and emission, and determine the potential energy surfaces for lattice vibrations. This information will allow us to determine equilibrium geometries, vibrational frequencies and the probabilities for intersystem crossing and radiationless decay. At present this requires an ab initio treatment, although other approaches may be useful if they can be shown to be sufficiently accurate.

The state of our ability to predict and analyze the optical spectra of impurity metals in crystals with ab initio methods is behind similar studies on molecules. Although ab initio molecular orbital methods were applied to this problem over 20 years ago by SUGANO and SHULMAN[1], current calculations continue to give large errors in the predicted energy level splittings. In the earlier calculations, limitations on the accuracy were imposed by small basis sets and the neglect of electron correlation. This was due to the size and speed of available computers and codes. In addition to the quality of the basis sets, other factors that need to be considered are the effects of the crystal lattice beyond the nearest neighbors, relativistic corrections, and spin-orbit interactions. We are currently including these factors in our theoretical calculations on transition metals in various crystal environments. Large basis set open shell Hartree-Fock (HF) calculations are being carried out on atom clusters to simulate the important effects of the crystal field and covalent bonding. A new configuration interaction code is being developed that will include the effects of both electron correlation and spin-orbit coupling[2]. The other important relativistic contributions are introduced by using an effective core potential derived from Dirac-Fock calculations on the free ion to replace the core electrons of the metal[3,4,5].

Recent experimental studies by McCLURE and co-workers[6] on Cu^+ in alkali halide crystals have provided a wealth of quantitative information to which we can compare our theoretical results. In the following sections calculations on Cu^+:NaF and Cu^+:NaCℓ clusters are discussed and compared to experiment and to recent X-alpha calculations.

II. Details of the Cluster Calculations

The cubic cluster for Cu^+ in NaF is shown in Fig. 1. The Cu^+ impurity is located at the center, the F^- ions are on the faces and at the corners of the cube, and the Na^+ ions are on the edges and on the axes outside the cube. All electrons on the Cu^+ ion were included in the calculations. The inner shell 1s electrons on the six nearest neighbor F^- ions were replaced by an effective core potential[7] and the eight F^- ions at the corners of the cube were approximated by point charges. The ten electrons on the Na^+ ions were also replaced by an effective core potential[8]. An external lattice potential was not included in this calculation, therefore the cluster has a net positive charge. From calculations on clusters of various sizes, this and the point charge approximation for the second shell of fluoride ions should not significantly change the relative location of the calculated energy levels.

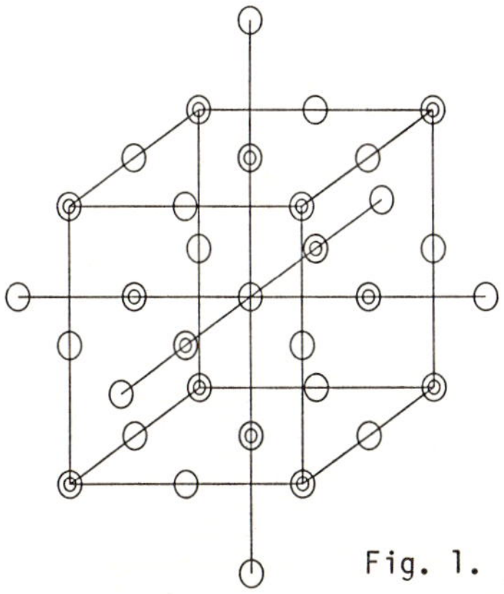

Fig. 1. The Cu^+:NaF cluster

The gaussian basis set for the fluoride ion is from KAHN et al.[7] augmented with a diffuse 2p function as recommended by DUNNING and HAY[9]. Comparison of the F^- basis set with and without this extra 2p function illustrates the important effects that even a small difference in the basis set can make. As shown in Table I, the calculated electron affinity of fluorine is greatly improved by the addition of the single diffuse function. Note that the best HF value in Table I is still far from the experimental value of 3.4[eV] [10]. BOTCH and DUNNING[11] have

Table I. Basis set effects on the energies of F and F^-

Basis set	[2s2p]		[2s2p] + 2p(0.074)	
Atom	F	F^-	F	F^-
E_{HF}(a.u.)	-23.9984	-24.0142	-23.9995	-24.0443
ϵ_{2s}	-1.5682	-1.0055	-1.5736	-1.0708
ϵ_{2p}	-0.7003	-0.1143	-0.7047	-0.1731
EA		-0.432 eV		-1.22 eV

Table II. Fluorine basis set effects on the atomic populations of the VF_6^{-4} cluster

Basis Set	Atom	s	p	d	Total
[2s2p]	V	6.16	12.38	3.70	22.24
	F	11.58	35.18	0.	46.76
[2s3p]	V	6.08	12.04	3.19	21.31
	F	11.61	36.08	0.	47.69

VF distance = 3.78[a.u.]

shown how the theoretical electron affinity can be systematically improved beyond the HF value, obtaining good agreement with experiment. Seriously underestimating the value of the electron affinity leads to an inaccurate charge distribution for the cluster and as a result the metal impurity will not experience the correct crystal field. An example of this is given in Table II. The doubly positive vanadium ion was surrounded by six fluoride ions with octahedral coordination. The VF distance wa fixed at 3.78[a.u.], close to the experimental value for V^{+2} in fluoride crystals Table II gives the electron populations on the atoms in the cluster for the two F basis sets from HF calculations on the $^4A_{2g}$ ground state.

The smaller basis set predicts that more than one electronic charge is transferred to the V^{+2} ion from the six F^- ions. By improving the F electron affinity with the extra 2p function the population on vanadium correctly corresponds to the electronic configuration $1s^22s^23s^22p^63p^63d^3$ of the ground state with only a reasonable amount of charge transfer.

The gaussian basis set for the Cu^+ ion was taken from WACHTERS[12]. Here again the accuracy of the calculation can be greatly improved by adding a single function and efficiently contracting the primitive gaussian basis set. HAY[13] has determined the exponent of a single gaussian 3d function to be added to the WACHTERS basis set and HOOD et al.[14] have shown how to flexibly contract the primitive basis set in addition to augmenting it with additional p and d functions. Table III illustrates the effects of varying the Cu^+ basis set. Hartree-Fock

Table III. Basis set effects on the energy levels of Cu^+

State	Basis I	Basis II	Numerical HF	Experiment
$3d^{10}4s\ ^2S$	-1638.8025a.u.	-1638.9033	-1638.9505	-
$3d^{10}\ ^1S$	6.23 eV	6.43 eV	6.41 eV	7.72 eV
$3d^94s\ ^3D$	8.63	7.76	7.69	10.53
$3d^94s\ ^1D$	9.16	8.19	-	10.97

Basis I (14s11p5d)/[8s6p2d] {62111111,511211,32}

Basis II (14s11p6d)/[10s9p3d] {5111111111,311111111,411}

calculations with two variations of WACHTERS' basis set were carried out on the ground state of neutral Cu and for the ground and excited states of Cu^+. The contraction for basis set I was taken from WACHTERS[12] and for basis set II from HOOD et al.[14]. The primitive sets are identical except that basis set II was augmented with an additional 3d function as suggested by HAY[13]. The results for each basis set are compared to the numerical HF calculations of FISCHER[15] and to experiment[16]. It is clear that the single additional 3d function has a dramatic effect on the excitation energies for Cu^+. Both basis sets are in reasonable agreement with the HF value for the ionization potential of Cu, however basis I overestimates the $3d^{10} \rightarrow 3d^94s$ excitation energy of Cu^+ by nearly an electron volt. Basis set II is in good agreement with the numerical HF value. Of course, due to the neglect of electron correlation differences in the ground (closed shell) and excited (open shell) states of Cu^+, the HF excitation energies are in error by $\sim$1.5 [eV].

III. Results for the Cu^+:NaF Cluster

Using basis set II for Cu^+ and the [2s3p] basis for F^-, the HF energies for the lowest five states were calculated for the cluster shown in Fig. 1. These are the $^1A_{1g}$ state which has a $3d^{10}$ Cu^+ configuration and the $^{1,3}E_g$ and $^{1,3}T_{2g}$ states which have a $3d^94s$ Cu^+ configuration. For these preliminary calculations, no basis functions were centered on the Na^+ ions. The effective core potentials describe the 10 core electrons of Na^+ and in addition to contributing to the crystal field for the Cu^+ ion, they provide the correct repulsive interaction with the electrons on the F^- ions. This is important for the determination of the equilibrium positions of the ions. By fixing all of the ions except the six nearest neighbor F^- ions at the NaF distance of 4.379[a.u.], the potential curves for the A_{1g} symmetric stretch motion of these ions were determined as shown in Fig. 2.

The curves show that the equilibrium CuF distance for the ground state is 4.62[a.u.] which is somewhat longer than for NaF. The excited state distances are 4.71[a.u.] for the ^3E_g and ^1E_g states and 4.72[a.u.] for the $^3T_{2g}$ and $^1T_{2g}$ states.

The vertical excitation energies are compared to the positions of the absorption bands measured by PAYNE et al.[17] in Table IV.

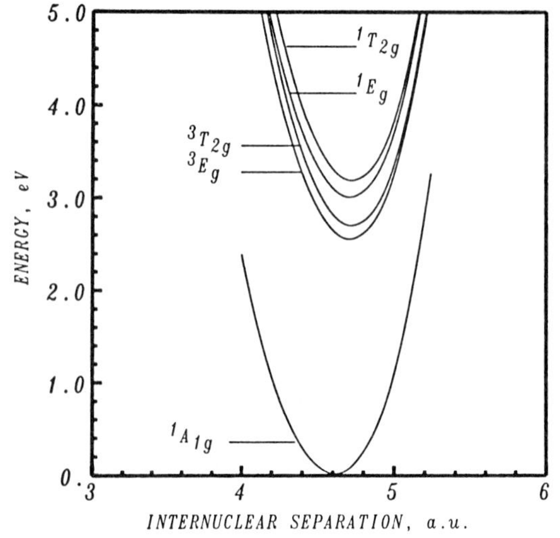

Fig. 2. Symmetric stretch potential energy curves for Cu^+:NaF

Table IV. Comparison of excitation energies for Cu^+:NaF to experiment

State	ΔE(cm^{-1}) calculated	ΔE(cm^{-1}) experiment[a]
$^1A_{1g}$	0.	
^3E_g	21224	31200
$^3T_{2g}$	22593	32600
^1E_g	24852	32200,34600[b],33500[c]
$^1T_{2g}$	26507	36400

a) Reference[17]

b) Jahn-Teller components from ref.[17].

c) Band peak, from average of Jahn-Teller components.

As was the case for Cu^+, the Cu^+:NaF calculations do not account for the electron correlation differences between the ground and excited states. Consequently the excited states are each about 10000[cm^{-1}] too low (the average of the Jahn-Teller components of the ^1E_g state is only 8648 [cm^{-1}] too low). The HF calculations on the excited states of the free Cu^+ gave correlation errors of ∿12000[cm^{-1}]. While placing the Cu^+ ion in the NaF host reduces the correlation error, both theory and experiment show that the excitation energies are increased by more than 1.0[eV]. The nearly constant difference between theory and experiment shown in Table IV is encouraging, since a uniform shift in the calculated levels will bring them into excellent agreement with the measured band peaks. This confirms the accuracy of the basis sets and the cluster approximation used here and suggests that when configuration interaction is included in future calculations good agreement with experiment will be obtained.

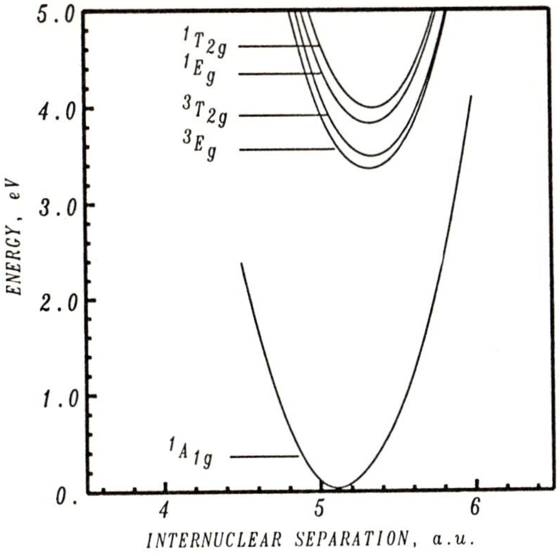

Fig. 3. The symmetric stretch otential energy curves for Cu^+:NaCℓ

IV. Results for the Cu^+:NaCℓ Cluster

By simply replacing the fluorine effective core potentials with those of HAY et al.[18] for chlorine and using the chlorine basis set suggested by KAHN[19] augmented with a diffuse p function with an exponent of 0.049 as determined by DUNNING and HAY[9], similar calculations have been carried out for Cu^+:NaCℓ. The symmetric stretch potential energy curves are shown in Fig. 3 which give the equilibrium CuCℓ distances for each state as $^1A_{1g}$, 5.11[a.u.], $^{1,3}E$, 5.33[a.u.], $^3T_{2g}$, 5.34[a.u.], and $^1T_{2g}$, 5.35[a.u.]. The ground state distance is shorter than the normal NaCℓ spacing of 5.31[a.u.] in contrast to Cu^+:NaF where the nearest F^- ions moved away from the Cu^+ion. The HF excitation energies are compared to the X-alpha calculations of CHERMETTE and PEDRINI[20] and the spectra of PAYNE et al.[21] in Table V.

Table V. Comparison of Hartree-Fock, X-alpha, and experimental excitation energies for Cu^+:NaCℓ[a]

State	HF	X-alpha[b]	Experiment[c]
^3E_g	3.59	3.61	
$^3T_{2g}$	3.77	4.13	
^1E_g	4.08	3.87	4.36,4.51[d]
$^1T_{2g}$	4.29	4.448	4.77

a) All units are in electron volts.
b) Reference[20]
c) Reference[21]
d) Estimated band peak from the average of the Jahn-Teller components.

The X-alpha excitation energies are for the calculated equilibrium ground state distance of 5.79[a.u.] while the HF values are for the CuCℓ distance of 5.11[a.u.] found from Fig. 3. Comparing the HF excitation energies for the singlet states to experiment, the ^1E_g state is too low by 3468[cm^{-1}] and the $^1T_{2g}$ state by 3871[cm^{-1}]. These errors are smaller than, but consistent with, the Cu^+:NaF calculations. It also suggests that the results for Cu^+:NaCℓ can also be brought into good agreement with the measured spectra by a nearly constant shift of all the levels together. The same is not true for the X-alpha calculations which give an error of 5161[cm^{-1}] for the ^1E_g state and 2339[cm^{-1}] for the $^1T_{2g}$ state. Also since these calculations do not directly determine the singlet and triplet states, they have reversed the order of the ^1E_g and $^3T_{2g}$ states. This would indicate that a great deal of caution must be exercised in using X-alpha results to interpret or predict the electronic spectra of impurities in crystals.

V. Summary

The present calculations on the Cu^+ ion in alkali halide crystals are preliminary in nature, but very encouraging. The relative positions of the excited states as determined by theory and experiment are in excellent agreement. The neglect of electron correlation appears to be the only major reason for the disagreement with the absolute location of the peaks of the absorption bands measured by McCLURE[6] and PAYNE et al.[17],[21]. Our future work will focus on including this and the effects of spin-orbit interaction as well as the influence of the remainder of the lattice.

Acknowledgements

Work performed under the auspices of the U. S. Department of Energy by the Lawrence Livermore National Laboratory under contract number W-7405-ENG-48. This work was supported by the Office of Basic Energy Sciences of the Department of Energy.

References

1. S. Sugano and R. G. Shulman:Phys. Rev. 130, 517 (1963).

2. R. M. Pitzer and N. W. Winter: "Polyatomic Calculations with Spin-Orbit Operators from Relativistic Core Potentials", presented at the American Conference on Theoretical Chemistry 1984, UCRL 90685A (1984).

3. W. C. Ermler, Y. S. Lee, K. S. Pitzer, and N. W. Winter: J. Chem. Phys. 69, 976 (1978).

4. P. A. Christiansen, K. S. Pitzer, Y. S. Lee, J. H. Yates, W. C. Ermler, and N. W. Winter: J. Chem. Phys. 75, 5410 (1981).

5. K. S. Pitzer: Int. J. Quantum Chem. XXV, 131 (1984).

6. D. S. McClure: "Cu^+ in Alkali Halides: Model Systems for Understanding Ions in Crystals", in Tunable Solid State Lasers, ed. by P. Hammerling, Springer Ser. Opt. Sci., Vol. 47 (Springer, Berlin, Heidelberg 1985)

7. L. R. Kahn, P. Baybutt, and D. G. Truhlar: J. Chem. Phys. 65, 3826 (1976).

8. S. Topiol, J. W. Moskowitz, C. F. Melius, M. D. Newton, and J. Jafri, "Ab Initio Effective Potentials for Atoms of the First Three Rows of the Periodic Table", ERDA Research and Development Report COO-3077-105, New York University (1976).

9. T. H. Dunning and P. J. Hay: "Gaussian Basis Sets for Molecular Calculations", in Modern Theoretical Chemistry, Vol. 3, ed. H. F. Schaefer III (Plenum, New York, 1976).

10. H. Hotop and C. Lineberger: J. Phys. Chem. Ref. Data 4, 539 (1975).

11. B. Botch and T. H. Dunning: J. Chem. Phys. 76, 6046 (1982).

12. A. J. H. Wachters: J. Chem. Phys. 52, 1033 (1970).

13. P. J. Hay: J. Chem. Phys. 66, 4377 (1977).

14. D. Hood, R. M. Pitzer, and H. F. Schaefer: J. Chem. Phys. 71, 705 (1979).

15. C. F. Fischer: J. Phys. B 10, 1241 (1977).

16. C. E. Moore: Atomic Energy Levels (Natl. Bur. Stand., Washington, DC, 1952) circular 467, Vol. 2.

17. S. Payne, A. B. Goldberg, and D. S. McClure: J. Chem. Phys. 78, 388 (1983).

18. P. J. Hay, W. R. Wadt, and L. R. Kahn: J. Chem. Phys. 68, 3059 (1978).

19. L. R. Kahn, private communication.

20. H. Chermette and C. Pedrini: J. Chem. Phys. 77, 2460 (1982).

21. S. Payne, A. B. Goldberg, and D. S. McClure: J. Chem. Phys. in press, (1984).

Cu^+ in Alkali Halides: Model Systems for Understanding Ions in Crystals

Donald S. McClure
Department of Chemistry, Princeton University, Princeton, NJ 08544, USA

The Cu^+ and Ag^+ ions at known lattice sites in alkali halide host crystals have been potentially simple systems to study as examples of electronically active impurity ions. However, the problems of interpreting their spectra were not solved until two-photon spectroscopy was applied. This article reviews the recent and current work using this method to assign absorption bands. In one especially favorable case NaF:Cu^+, it was possible to determine the quantum levels of the t_{1u} mode in which the Cu^+ vibrates off the lattice site, by using high resolution one- and two-photon spectroscopy. This led to the construction of potential curves for this important mode in several excited electronic states. Other experiments including magnetic effects in triplet states are described. Theoretical interpretations at various levels are also reviewed.

I. Introduction

Among the simplest electronically-active metal ions which can be studied in host crystals are Cu^+, Ag^+ and Au^+. They have d^{10} closed shells ($d^4_{3/2}\ d^6_{5/2}$ for Au^+). The first excited states are from the d^9s configuration, and for Cu^+ some states of $3d^94p$ have been identified in crystals. These systems are especially useful as models of impurity ions in crystals since their spectra are relatively simple, yet have enough features to be informative. Through a combination of one- and two-photon spectroscopies and theory, one can reach a quite detailed understanding of these systems.

There has been a long history of research on Cu^+ and Ag^+ in simple crystals such as the alkali halides. [1] Some of the early work showed that the near UV bands are due to the $d^{10}\rightarrow d^9s$ transitions, [2] but they occur up to 2eV higher in energy than the corresponding transitions in the free atom. [3] The $d^{10}\rightarrow d^9p$ transitions occur in the vacuum UV, 1-2 eV below the free-ion transition energy. [3,4]

II. Crystal Production

The most significant new information on the spectroscopy of Cu^+ in crystals has come from NaF:Cu^+. [5,6] Since CuF does not exist at room temperature there was a question as to whether NaF:Cu^+ could exist. The instability of CuF arises from the process $2CuF \rightarrow Cu + CuF_2$. [7] At high temperatures the equilibrium shifts to the left, increasing the amount of CuF. If one assumes that CuF has the same lattice energy as NaF then the above equilibrium is exothermic by 178 kJ/mole of CuF at 300K. Application of the Gibbs-Helmholz equation shows that there will be a large shift in the equilibrium constant favoring CuF on going from 300K to 1291K, the melting point of NaF. At the melting point of KF, 1156K, the equilibrium is a factor of 11 less favorable to CuF. For LiF and RbF the calculated equilibrium point is less favorable still. Samples of NaF:Cu^+ from H. BILL (Geneva) [8] and De LONG (Utah) had Cu in its three valence states but substantial fractions were Cu^+, probably about 100 ppm.

In the case of KF:Cu^+, our estimate for a sample grown by De LONG was 5 ppm.

A sample of RbF:Cu grown by De LONG contained small amounts of Cu^+, but in many different states [9]. Samples of LiF grown with Cu had no detectable Cu^+.

There is no problem in growing Cu^+-containing alkali chlorides, bromides and iodides if reducing conditions are maintained as in a graphite crucible.

III. The Ground State of Cu^+ in Alkali Halides

The ionic radius of Cu^+ appears to be about the same as for Na^+; however, Cu^+ substitutes for any of the alkali metals in alkali halide crystals, usually at a single unique site. This site is not always the center of the halide octahedron; in fact, in the majority of cases, the Cu^+ is displaced from the center.

One important source of information about the ground state equilibrium position is the oscillator strength of the $^1A_{1g} \rightarrow {}^1E_g, {}^1T_{2g}$ transitions. If the Cu^+ lies at the center of inversion at the alkali metal site these transitions are symmetry-forbidden, and are thus weak and follow a temperature dependence of the form

$$f(T) = f(0) \coth \hbar\omega/2kT \qquad (1)$$

NaF:Cu^+ follows this form with $\hbar\omega = 200\ cm^{-1}$; for LiCl:$Cu^+$, $\hbar\omega = 100\ cm^{-1}$.

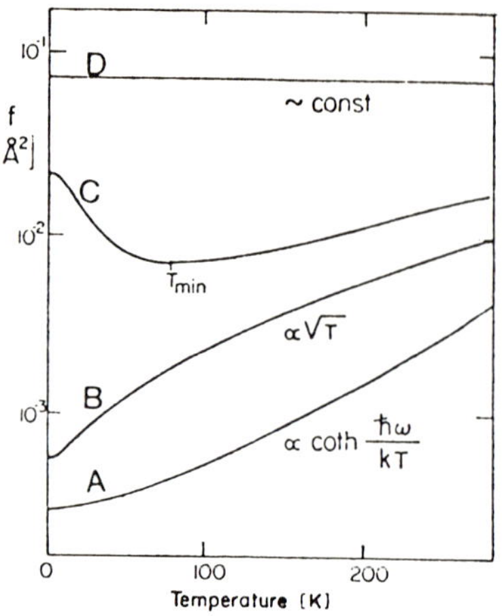

Fig. 1. Schematic display of four types of temperature dependence for absorption strength of Cu^+ in alkali halide host crystals, after FUSSGANGER [10].

These frequencies are within the band of optical modes of the lattice, and thus one can assume that the impurity is strongly bound to the center of the halide octahedron.

In the case of KCl:Cu^+ the transition is stronger and less temperature-dependent than in the two foregoing examples. Here the Cu^+ is off-center and the center of inversion selection rule does not apply.

FUSSGANGER has examined many examples of the temperature-dependence of the oscillator strength and has been able to classify them into four categories, as shown in Fig. 1. [10, 11] Between the two extremes of the coth behavior and independence of temperature are the weakly centered, anharmonic oscillator case, such as NaCl:Cu^+, and the examples in which the Cu^+ goes from an off-center site to an average on-center site at higher temperatures. Figure 1 shows these temperature dependencies.

Later work by LUTY and his coworkers combined dielectric and optical monitoring with the application of hydrostatic pressure.[12] They were able to make quantitative estimates of barrier heights and displacements.

SIEVERS and his coworkers have shown that low frequencies in the infrared and Raman are associated with vibration at weakly bound sites (e.g. NaCl:Cu^+ where ω = 23 cm^{-1}) and with tunneling motions between equivalent off-centered sites as in KCl:Cu^+. [13]

IV. Excited States of Cu^+ and Ag^+

The excited states of Cu^+ and Ag^+ in crystals are closely related to the states of the d^9s and d^9p configurations of free Cu^+ and Ag^+, as proposed by SEITZ. [2] Figure 2 shows these relationships for NaF:Cu^+ and NaCl:Ag^+ by

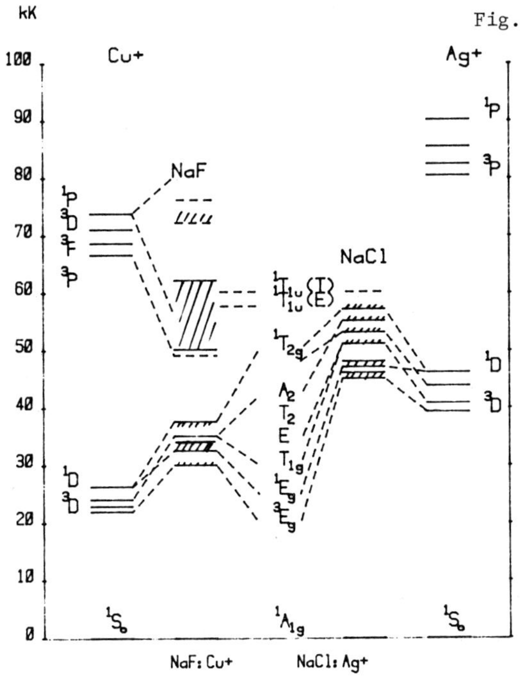

Fig. 2. Energy levels of Cu^+ (left) and Ag^+ (right) as free ion (outer levels) and in NaF (Cu^+) or NaCl (Ag^+) (inner levels). The band edges of the host crystal shown are the practical transparency limits in a 1 mm crystal.

means of energy-level diagrams. The impurity levels shown in this figure are from recent experimental work. [5,6,14] The lower edge of a cross-hatched band is our best estimate of a zero phonon energy for the given electronic state, and the width corresponds to the origin to peak energy. It is possible to show a nearly complete 1:1 correspondence between the spectral bands of Cu^+ and Ag^+ in a host crystal since most of the triplet components have now been observed and assigned by two-photon spectroscopy. [5,14] These components are $^3E_g(T_{1g},T_{2g})$, $^3T_{2g}(A_{2g}, E_g, T_{1g}, T_{2g})$.

The d^9p states are only partially known. The vacuum UV bands of LiCl:Cu^+, [4] NaF:Cu^+[6] and NaCl:Cu^+ [15] are directly measurable as distinct bands (except that only half of the NaCl:Cu^+ band is seen). The vacuum UV band of KF:Cu^+ has also been observed, [9] and indeed is one of the best proofs that Cu^+ is present since the amount of Cu^+ in these crystals is too small for the near UV band to be clearly seen. Since two $^1T_{1u}$ states can be derived from d^9p one might expect to see two bands. But these states are separated by only 10Dq = 3000 cm^{-1}, and the bands are much broader than this. In NaF:Cu^+ there is some evidence for two bands at low temperatures. There are four other singlet states in this configuration, but they are not expected to be identified because they are forbidden in O_h symmetry, and the band widths are too great. For the other alkali halides the exciton band interferes with observations of the Cu^+ d→p transition.

The triplet states of Cu^+ may give bands of appreciable intensity as seen in the near UV spectra. There is no clear evidence for them in one-photon vacuum UV spectra, however. We expected to see evidence for them at the long wavelength end of the vacuum UV band since the triplets should be some 3000 cm^{-1} lower than the singlets. (In the free ion $^3P_1$ lies 5679 cm^{-1} below $^1P_1$). In recent work, RING-LING CHIEN has measured the excited state absorption bands originating on the $^3E_g(d^9s)$ state for Cu^+ in NaF, LiCl, NaCl and KCl. [16] Transitions from the two close-lying components of ^3E_g, T_1 and T_2 can end on any of the 19 triplet components in d^9p, although we do not know the relative transition probabilities at this time. The results of these measurements are that the triplet-triplet absorptions are very broad bands which, when the origin shift is accounted for,lie right on top of the singlet-singlet bands. Thus, there is no perceptible shift between singlets and triplets of d^9p within the broad limits imposed by the band-widths. This result probably shows that the p-electron, d-hole exchange interaction is even smaller in the crystal than in the free ion.

The assignments of the bands in NaCl:Ag^+ were made by two-photon spectroscopy [14,17]. They were clearly separable into E_g and T_{2g} classes, and four of the triplet components were identified. Since these bands extend out to 1700 Å and include all bands before the cutoff of NaCl, there are no contributions from the $4d^95p$ configuration. In the past, the higher bands had been assigned to it. Thus the 5p levels, which lie near 90,000 cm^{-1} in the free ion, are not lowered enough in NaCl to contribute to the spectrum of Ag^+. The spectrum of Ag^+ in NaCl is confusing because the T_2 component of $^3T_2$ is almost as strong as the ^1E_g band, and the $^1T_2$ band is many times stronger than the ^1E_g band, thus lending credence to its misassignment to a 5p band.

V. One- and Two-Photon Spectra of NaF:Cu^+

Since NaF is transparent to about 1300 Å, the one-photon spectra of Cu^+ can be studied over a range of 76000 cm^{-1}, or 9.5 eV. Figures 3 and 4 show the near and vacuum UV spectra of NaF:Cu^+ at about 4K and 300K.[5,7] These represent the 3d→4s and 3d→4p transitions, respectively. In the VUV, only the band at 60,000 cm^{-1} can be definitely attributed to Cu^+; the 70,000 cm^{-1} band when excited with synchrotron radiation does not lead to the Cu^+ luminescence, while the lower energy band does. [18]

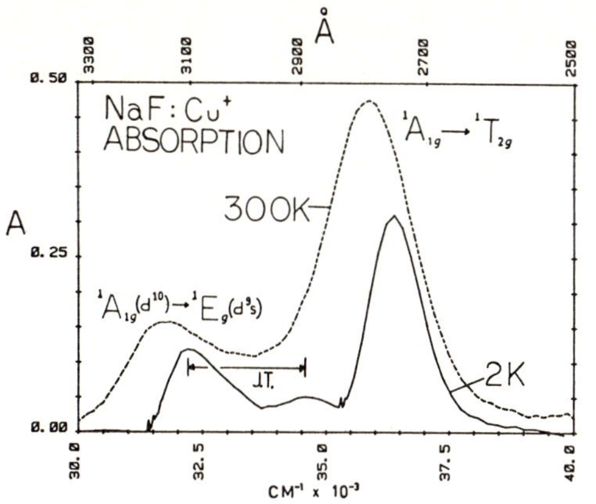

Fig. 3. The ${}^{1}A_{1g}(d^{10}) \rightarrow {}^{1}E_{g}$, ${}^{1}T_{2g}(d^{9}p)$ transitions of Cu^{+} in NaF at 2K and 300 K. [5]

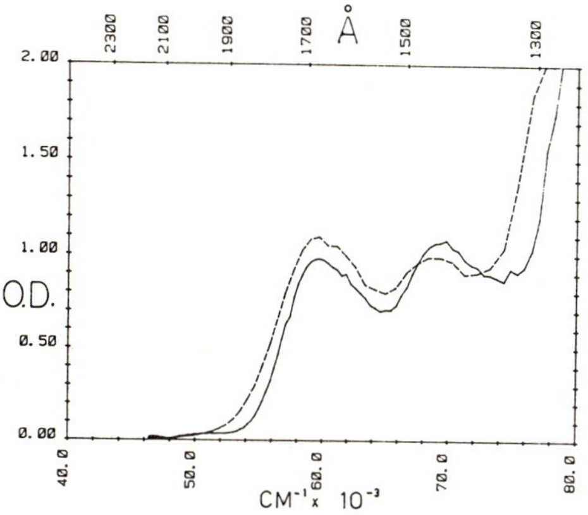

Fig. 4. The ${}^{1}A_{1g}(d^{10}) \rightarrow {}^{1}T_{1u}(d^{9}s)$ transitions of Cu^{+} in NaF at 10 K and 300 K (dashed line).

In the two-photon absorption in a cubic crystal, two polarizations are possible, and thus more selection rules are available than for one-photon absorption. In the simplest arrangement, both photons have the same frequency and polarization, but the polarization is rotated from the 100 axis toward the 110 axis, when the light is propagated along 001. The polarization rules have been worked out for cubic crystals by Bader and Gold, [19] and for the case at hand they give the transition probabilities for the ${}^{1}A_{1g} \rightarrow {}^{1}E_{g}$ and ${}^{1}A_{1g} \rightarrow {}^{1}T_{2g}$ transitions of Cu^{+} in a Na^{+} site of NaF:

$$P_E = 1 - \frac{3}{4}\sin^2 2\theta, \quad P_T = \frac{3}{4}\sin^2 2\theta \qquad (2)$$

where θ is the angle between the 100 axis and the polarization vector.

The two-photon spectra were obtained with a Molectron DLII dye laser pumped by a UV-24 nitrogen laser. The fluorescence of the crystal detected by a photomultiplier gives the magnitude of the two-photon absorption.

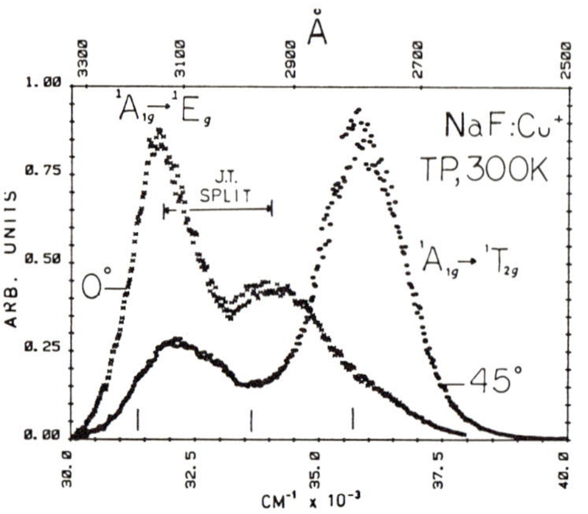

Fig. 5. The two photon spectrum to Fig. 3 at 300 K, showing 0° and 45° polarizations.

The two-photon excitation spectra of $NaF:Cu^+$ at 300K are shown in Fig. 5. When compared to Fig. 3, one can see how much more information has been revealed by this two-photon method. The E-band clearly shows the double-humped feature due to the Jahn-Teller effect, and it is clearly separated from the T_2 band. If one subtracts 1/4 of the $\theta=0$ spectrum from the $\theta=\pi/4$ spectrum, the unmixed T_2 spectrum is obtained, as shown by the selection rules. Not only is the $^1T_2$ band given its correct shape, but a new band is found: this is probably the T_2 component of the $^3T_2$ state, since it has the correct position and intensity for such an assignment. [5]

The excitation spectrum of the ^3E state has been measured by a one-photon laser excitation experiment, since it lies at longer wavelengths than the ^1E band. [20] The luminescence of Cu^+ comes entirely from this state. The luminescence spectrum of $NaF:Cu^+$ peaks at 3700 Å. The lifetime of the luminescence shows a remarkable temperature-dependence shown in Fig. 6

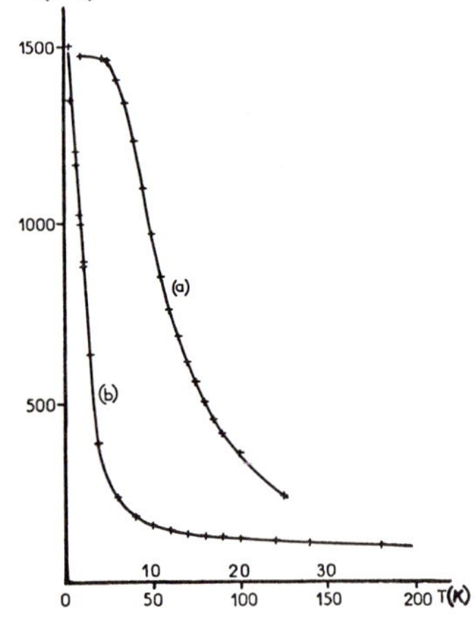

Fig. 6. The triplet state lifetimes of $NaF:Cu^+$ as a function of temperature. Note that curve (a) is to be read from the upper temperature scale and curve (b) from the lower. [MOINE and PEDRINI, 21].

without any change of quantum yield. [21] The reason for this will be given in Section VII.

The low temperature one-photon spectrum of NaF:Cu^+ shows fine structure at 3179 and 2838Å. Under higher resolution additional structure can be seen. Since the transition is g→g, these cannot be zero phonon lines, but must represent the enabling mode which breaks the parity. Their interpretation is made clear when the two-photon spectrum is observed. Transitions g→g now become allowed and the true electronic origin (zero phonon line) is observed. Figure 7 shows the fine structure in both types of spectra for both singlet states. We see here a progression in a t_{1u} (off-center) mode of the Cu^+, the even quanta appearing in the two-photon spectrum, and the odd quanta in the one-photon spectrum. In both states the progression is highly anharmonic. The analysis of these features leads to the potential functions for the off-center motion of the Cu^+, discussed in Section VI.

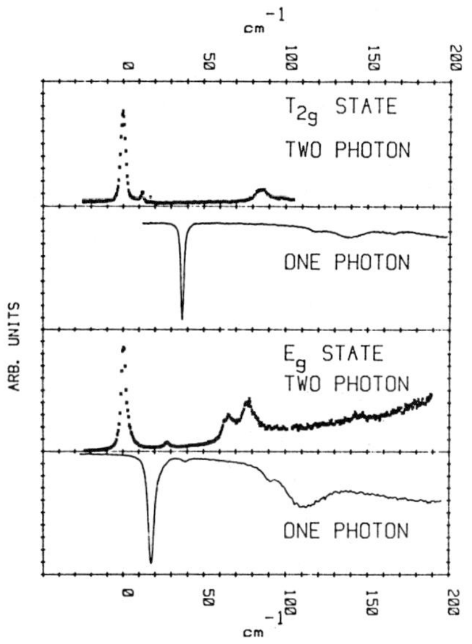

Fig. 7. One-and two-photon laser excitation spectra near the origins of the $^1A_{1g} \rightarrow {}^1E_g$ and $^1A_{1g} \rightarrow {}^1T_{2g}$ bands of NaF:Cu^+ at 2K. The $^1T_{2g}$ origin labelled 0 is at 35333 cm^{-1} and the ^1E_g state origin is 31433 cm^{-1}.

Further spectroscopic study of the Cu^+ spectrum should reveal details of the two triplet states. It would be interesting to compare their potential functions to those of the singlets. In work currently in progress a nearly harmonic progression in a very low frequency has been observed in the laser excitation spectrum of the ^3E_g state. [20] The two-photon spectrum of the $^3T_{2g}$ state can be found by using two perpendicularly polarized photons; there should be no interference from the ^1E_g two-photon spectrum in this case. [19]

VI. Vibrational Levels and Potentials in the Electronic States of Cu^+ in NaF

By various methods it was possible to measure or estimate the values of the frequencies of three vibrational modes in three electronic states for NaF:Cu^+. [5] These results are given in Table 1.

TABLE I. Vibrational frequencies of NaF:Cu^+ (cm^{-1}).

Electronic state	Vibration a_{1g}	e_g	t_{1u}
$^1A_{1g}$	150	150	200
^1E_g	150	300	17.5
$^1T_{2g}$	150	---	36

The t_{1u} mode in the ground state was determined from the plot of intensity vs temperature which was found to obey the coth $\omega/2kT$ law with $\omega = 200\pm10$ cm^{-1}. The Cu^+ is held firmly at the center of the octahedron of F^- ions in the ground state: it is "on-center".

We must now consider the remarkable fine structure seen in the excited singlet states and shown in Fig. 7. The assignment of this fine structure to the t_{1u} off-center motion of Cu^+ in the NaF lattice is compelling, even though there has been a drop from 200 cm^{-1} in the ground state to 17 and 36 cm^{-1} in the excited states. The reason for such a large change is of great importance and will be discussed after the analysis of the spectra.

The progressions in the $^1T_{2g}$ and ^1E_g states are strongly anharmonic and the higher members are split as well as being broadened. To get started, we fitted these progressions to one-dimensional oscillator models and found that the $^1T_{2g}$ progression fits a quartic oscillator (when 117, 138 and 166 are averaged together for the v=3 level). The ^1E_g progression fitted a double well potential model. [5]

To get a more realistic model, we began with a three-dimensional harmonic oscillator basis set and added a negative harmonic term, a positive quartic term and an angle-dependent crystal field term. [22] The negative harmonic term provides the central maximum in the potential. The crystal field term determines whether the force constant in the direction of the F^- ions is greater or smaller than in intermediate directions. Our procedure was to solve the problem of the spherical oscillator (including the radial potential terms only) including 200 levels. The lowest eigenvalues and eigenfunctions were used in the calculation of the crystal field energies and Franck-Condon factors.

The radial Hamiltonian is thus

$$H = H^{HO} + r^4 + \alpha r^2 \qquad (3)$$

where α may be negative, and $\alpha+0.5$ is the actual coefficient of r^2 (in dimensionless form). The harmonic oscillator term, H^{HO}, provides the basis set. We followed the procedure of BELL et al. [23] to generate the matrix elements. The quantum numbers are a radial vibrational, V, and an angular momentum, ℓ. The selection rules are $\Delta\ell=0$ and there are no matrix elements between even and odd V. The center line of Fig. 8 shows the nature of the purely radial solution. The example is for the case $\alpha=-1.1$ which approximately fits the $\ell=0$ and $\ell=1$ levels of the $^1T_{2g}$ state. Levels with $\ell>2$ are split in the first order by the crystal field.

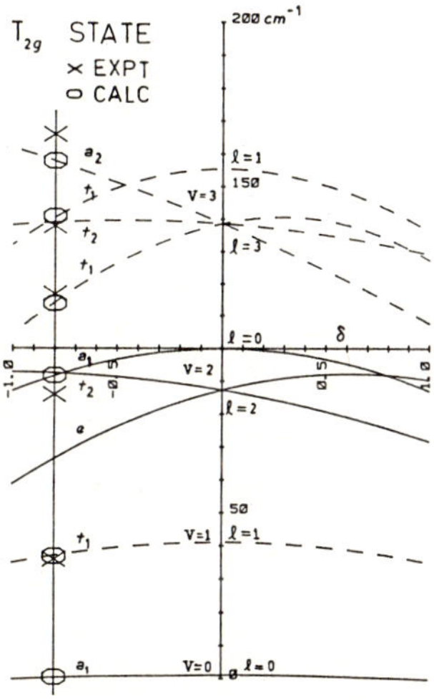

Fig. 8. Calculated energy levels for Cu^+ t_{1u} mode in the $^1T_{2g}$ state up to V=3 quantum level. Center lines give spherical oscillator levels labelled by V and ℓ (angular momentum). The crystal field splitting of these levels is given by δ. Observed levels are fitted at the left with $\delta=-0.8$.

The crystal field contains potential terms belonging to L=0,4 when expressed in terms of spherical harmonics. The Hamiltonian becomes

$$H = H^{HO} + \alpha r^2 + r^4 + \delta[C_4^0 + \sqrt{\frac{5}{14}}\,(C_4^4 + C_4^{-4})]r^4 \qquad (4)$$

where δ is the vibrational crystal field parameter. This equation was not solved with high precision: the crystal field was used as a perturbation on the levels of the spherical oscillator, and only the matrix elements between a V-level and V±2, V±4 were considered. This procedure seemed ade-

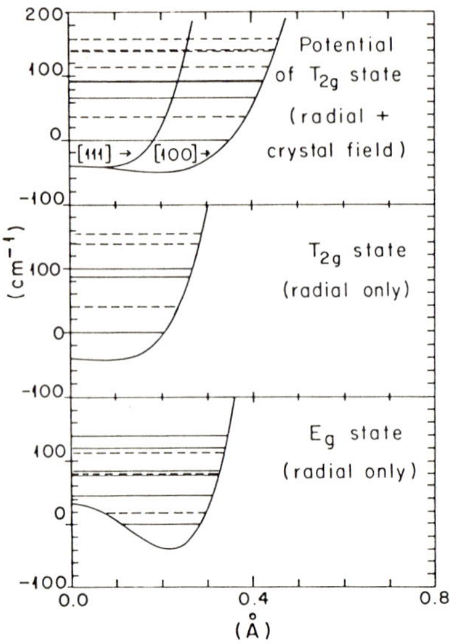

Fig. 9. Potential surfaces and the corresponding energy levels for: top, the $^1T_{2g}$ state with α=-1.1, δ=-0.8 showing cuts along [111] and [100] axes ; middle, the radial part of the above potential; bottom, the radial part of the ^1E_g potential.
--- one photon
—— two photon

quate for the $^1T_{2g}$ state, but it was not for the ^1E_g state. Figure 8 shows the results of this calculation for the $^1T_{2g}$ state, where δ varies between 1 and -1. The splitting of states having $\ell \geqslant 2$ is shown to be quite large relative to the spacing of the unperturbed levels in the vicinity of $|\delta|$=1. It was possible to fit the five intervals of the $^1T_{2g}$ state to the computed levels when δ = -0.8, α = -1.1, and when a scaling factor of 21.3 cm^{-1} was used as shown in Fig. 8. The scaling factor was determined so as to make the 0-1 interval of the dimensionless calculated spectrum match the 0-1 interval of the observed spectrum. The top panel of Fig. 9 shows the poten-for the $^1T_{2g}$ state.

The same scaling factor was used to fit the ^1E_g intervals. Although the crystal field perturbation was too large for the limited basis size we used, the ℓ=0 and ℓ=1 levels of ^1E_g can be fitted by the spherical oscillator potential when α=-4.2. Although such a fit may misassign some levels, we can get part of the physical picture of the potential for the ^1E_g state. The central maximum is 73 cm^{-1} and two levels lie below it, as shown in the lower panel of Fig. 9. The radial potential for the $^1T_{2g}$ state shows only a 1.5 cm^{-1} central barrier and all levels lie above it (center panel).

The top panel of Fig. 9 shows two cuts through the potential surface at the 100 and 111 axes for the $^1T_{2g}$ potential. This plot shows that the potential is lower when the displacements are along the Cu-F direction than when they are along the intermediate directions.

The major factor which affects the potential for the t_{1u} mode in any of these excited states is the replacement of a d-electron by an s-electron. We will see in sec. IX that there is a large negative harmonic potential term because the 4s-4p mixing is a strong function of the displacement from the center. It is probable that the $^1T_{2g}$ state which interacts with the ligands via π-antibonding has a smaller anisotropy than the ^1E_g state which interacts via σ-antibonding.

The triplet components, of which three have been observed, should have t_{1u} potentials similar to those found for the singlets. There is a good chance that by using two-photon methods the fine structure can be found and then analyzed to reveal these potentials. The data for all these d^9s states would constitute a unique collection which should be taken very seriously by theorists.

VII. The ^3E_g State in Absorption and Emission.

The remarkable dependence of emission lifetime on temperature shown in Fig. 6 for the case of NaF:Cu^+ is a general feature of the Cu^+ emission in any host. PEDRINI interpreted this fact by showing that the two components T_1 and T_2 of ^3E_g have different fractions of singlet character, and the one with the least singlet character, T_1, lies lower. [24,25] Thus, at low temperatures, only a slow emission is possible, but at higher temperatures the T_2 state may become occupied, leading to more rapid emission. Pedrini's analysis of the kinetics showed that the splitting between these states is on the order of 50 cm^{-1}.

The T_1 and T_2 components of ^3E_g are mixed by a magnetic field. Therefore, it was possible to test Pedrini's hypothesis by studying the effect of a magnetic field on the lifetime at a low temperature. The magnetic effects were indeed found and gave values of the splitting in substantial agreement with those obtained from the kinetics. [25] In the case of NaF:Cu we found Δ= 26 cm^{-1}, while Pedrini found Δ= 27 cm^{-1}. For LiCl:Cu^+ we found Δ= 60 cm^{-1} while the kinetic analysis gave 80 cm^{-1}.

The T_1-T_2 splitting is a complicated function of the spin orbit parameters and of the energy separation from the singlets. It will, however, be a useful piece of data in combination with other measurables in a complete theory of the ion in the crystal site

The ^3E_g triplet should have a Jahn-Teller distortion in the subspace of the e_ε and e_θ vibrational modes at the impurity site, and a distortion in the subspace of the t_{1u} off-center mode as in the corresponding ^1E_g state. With the discovery of fine structure due to the t_{1u} mode in the one-photon spectrum [20] there is a good chance that a t_{1u} potential function for the triplet can be worked out. If the two photon spectrum can be observed this function could surely be obtained. Laser excitation in the one-photon mode over the broad band has not revealed a double humped peak which would give the Jahn-Teller energy, but there is a chance that it will be found in the two-photon spectrum. [20]

VIII. Two-Photon Spectra of Other Cu^+-Alkali Halide Systems: Vibronic Mixing of States

Polarized two-photon spectra of LiCl:Cu^+, NaCl:Cu^+, KCl:Cu^+ and NaCl:Ag^+ have been studied [26] and are shown in Fig. 10. The first two of these have spectra at 300K, similar to that of NaF:Cu^+ (Fig. 5). The same system, using two parallel photons was used as in the NaF:Cu^+ experiments. When subtracting a quarter of the 0° spectrum from the 45° spectrum to generate the pure T_2 spectrum, we found that there appeared to be too much long wavelength absorption. In the case of NaF:Cu^+ this region was attributed to triplet state absorption, since it was a distinct band and had a reasonable intensity ratio to the main T_2 band. The NaCl:Cu^+ and LiCl:Cu^+ cases, however, appear to suggest depolarization by some process. An extreme case of depolarization is shown by KCl:Cu^+.

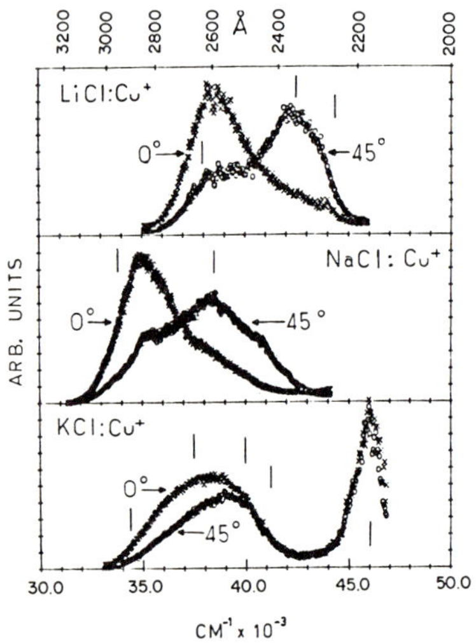

Fig. 10. Two photon spectra for LiCl:Cu^+, NaCl:Cu^+ and KCl:Cu^+ at 300 K.

The conclusion was that depolarization is caused by mixing of E_g and T_{2g} states either through vibrational displacements in the excited state or by static off-center displacements which mix these two symmetries. The Franck-Condon Principle shows why ground-state displacements are carried to the upper states where they contribute to this mixing.

A theory of the mixing is based on an expansion of the Hamiltonian in terms of the t_{1u} off-center vibrational displacements. The first term which can mix E_g and T_{2g} states is a second order one having t_2 symmetry:

$$V_{t_2} = \Sigma \frac{\partial^2 H}{\partial Q_x Q_y} \quad Q_x Q_y = A(Q_x Q_y + Q_y Q_z + Q_z Q_x) \qquad (5)$$

where the Q_x etc. are components of the t_{1u} mode. This perturbation provides off-diagonal matrix elements between the partners of the E_g and T_{2g} states, mixing their wavefunctions and thus mixing the two-photon polarization properties. The mixing is proportional to the fourth power of the displacement.

This depolarization effect could not have been identified in a one-photon spectrum. Such spectra simply present a composite of the two states with no characterization.

The method of doing two-photon spectroscopy with two independently polarized beams of different frequency would be an improvement over the present experiments because the T and E states would be more cleanly separated.

IX. Theory of the Ion in a Crystal: $NaF:Cu^+$

The many pieces of experimental data collected during the course of this research deserve a unified-theoretical framework, which one hopes may be provided some day. Here, however, we can present some simple calculations which when compared to the data show some of the shape of a more complete treatment.

The value of 10Dq, the d-hole splitting factor is quite small compared to the values for divalent ions. We were able to extract this value from the data after correcting for the Jahn-Teller distortion of the ^1E_g band, using the two-photon spectra. Its value for $NaF:Cu^+$ is 1800 ± 100 cm^{-1}. The value of 10 Dq was not known for any Cu^+ species before this work since only by two-photon spectroscopy is it possible to separate ^1E_g from $^1T_{2g}$ and to determine the Jahn-Teller splitting.

A calculation of Dq was made for Cu^+ in NaF, NaCl and LiCl using the DVXα program of Ellis, and making the Cu-X distance the same as the M-X distances [26]. The Dq values obtained were about 50 to 100 percent higher than the experimental value. The e_g orbital was 88 percent 3d, 12 percent F^- pσ for the case of NaF; the t_{2g} orbital was 97 percent 3d and the a_{1g} orbital 92 percent 4s. The chloride orbitals were slightly more mixed. The case of $NaCl:Ag^+$ was also calculated, and it showed much more orbital mixing than for Cu^+, consistent with a Dq value 4-5 times greater.

These are reasonable results. The s-electron of Cu^+ is antibonding to the ligands but remains relatively pure. The $d\pi$ orbitals are even less mixed with ligand orbitals.

A more complete calculation was made by HARRISON and LIN [27] for the case of LiCl:Cu^+ using the L.C.A.O. method starting with a self-consistent field for LiCl. They were able to obtain a description of the 4s and 4p orbitals of Cu^+. The 4s orbital they obtained was very similar to the less complete DVXα calculation. The greater antibonding of 4s to ligand $p\sigma$ orbitals compared to the 3d orbitals accounts for its higher excitation energy in the crystal compared to the free ion. The 4p orbital is not as localized as the 4s orbital and mixes with conduction band states. Since these are generally higher than the p-electron energy, the energy of the latter is reduced in the crystal. These results for d, s and p electrons explain the main features of Fig. 2.

It would be interesting to apply these theoretical methods to the excited state potential functions to see what the important factors are. These methods become much more complex when symmetry is reduced, however, and we have instead done some simpler calculations. One important effect of the off-center displacement of Cu^+ in the F_6^- octahedron of NaF is that 4s and 4p orbitals may mix. We carried out a perturbation calculation for the energy reduction of the displaced 4s orbital assuming that the 4p band at 1700 Å represented all of the 3d→4p oscillator strength. [22]. This assumption is supported by HARRISON and LIN's calculation for LiCl:Cu^+.[27] The energy reduction of the excited state due to s-p mixing is:

$$E(\rho) = \frac{\langle 4s|\frac{\partial H}{\partial \rho}|4p\rangle^2}{E_s-E_p}\rho^2 \qquad (6)$$

where ρ is the off-center displacement. The vibronic coupling matrix element can be evaluated if it is assumed that the oscillator strength of the d→s band is contributed entirely by the d→p band.

$$f_{d\to s} = f_{d\to p}\frac{\langle 4s|\frac{\partial H}{\partial \rho}|4p\rangle^2}{(E_s-E_p)^2}\langle\rho_o\rangle^2 \qquad (7)$$

where $\langle\rho_o\rangle^2$ is the rms ground state displacement in the t_{1u} mode. Thus we have

$$E(\rho) = (E_s-E_p)\frac{f_{d\to s}}{f_{d\to p}}\frac{\rho^2}{\langle\rho_o\rangle^2} \qquad (8)$$

The data for using this result are available and one finds $E(\rho) = -3.7\rho^2$ with -3.7x2 md/Å being the negative force constant, (ρ in Å). This value is higher than the ground state harmonic oscillator force constant of 1.4 md/Å: thus one can say that there is plenty of force from the s-p mixing to drive the Cu^+ ion off-center in the d^9s states, but that it is much larger than is observed. The value of α in the potential function fitted to the t_{1u} mode in the $^1T_{2g}$ state is only -0.07 md/Å. Therefore, there are important quadratic forces opposing the off-center s-p force. The s-p force may also be over estimated by this calculation. Seeing these results, one wishes for a better calculation, but it is easy to see that the off-center potential may be an important feature of the excited state for Cu^+ or Ag^+ in any host crystal, since the basic s-p mixing, according to the above calculation will not differ much from one system to another. It is therefore safe to assume that even when the fine structure cannot be observed, the excited state will have an appreciable off-center component in its t_{1u} potential.

The Jahn-Teller potential surface was partially revealed by the two photon experiments, and it too is expected to have about the same features in any host crystal. Some details of the interaction between the Jahn-Teller distortion and the t_{1u} off-center distortion were observable in the case of NaF:Cu^+. [22]

X. Conclusions

The value of two-photon spectroscopy lies in the polarization choices available even for cubic crystals, and in its ability to excite transitions between states of the same parity. The majority of electronically active ions in crystals have same-parity transitions throughout the near IR and visible. The zero phonon line and its sidebands obtainable for these cases by two-photon methods are of prime importance for understanding the excited states. Separating states of different symmetry properties where they partially overlap is another of the important uses of two photon spectroscopy.

The Cu^+ and Ag^+ ions in crystals are ideal candidates for study by two-photon spectroscopy using present day laser technology: the dyes need only work in the convenient visible region in order to reach the first absorptions in the UV, and the fluorescence used for detection appears in the near UV at a considerable wavelength separation from the laser light from which it is easily separated. It would be very helpful if the types of results obtained here could be obtained for any ion in a crystal host. Entirely different methods will have to be developed, however, since transition metal ions do not have such convenient fluorescences and the interesting spectra occur in visible and near infrared.

Another point to be made, already well known but worth repeating, is the value of a host crystal having light atoms - fluoride rather than chloride for example - in order to reduce the S-factor and make zero phonon lines more detectable. The NaF:Cu^+ example is so far the only one of the alkali halides with Cu^+ or Ag^+ showing fine structure. The good radius match between Na^+ and Cu^+ and favorable thermodynamic properties are also factors in this case.

If we consider the things we want to know about the physics of impurity ions in crystals, the examples of Cu^+ in alkali halide hosts have been very educational. It is unusual to be able to observe the behavior of electrons in three different atomic orbitals in one system, yet in Cu^+ the 3d, 4s and 4p orbitals contribute to its observable properties. LIN's work on LiCl:Cu^+ has been a good start on the theoretical desciption of these electronic properties. [27]

Next, the unusual amenability of the spectra of Cu^+ to study by two-photon spectroscopy made it relatively easy to get data leading to detailed knowledge of the dynamics of the ion in its crystal site in several electronic states. This kind of information is needed in order to reach a thorough understanding of the radiationless processes between states. From the example of Cu^+ in NaF we can get some impression of the possible complexity of the potential surfaces of excited electronic levels.

In the ^1E_g state, for example, we were able to measure the 2250 cm^{-1} e_g Jahn-Teller splitting, the off-center potential in the t_{1u} mode leading to a central maximum of 73 cm^{-1}, and also were able to measure quantitatively some results of the anharmonic coupling between the e_g and t_{1u} modes.

The NaF:Cu^+ system will yield further detailed information when new types of two-photon studies on the triplet states are completed. Line shape studies using narrow line lasers will add further information on radiationless transitions and lattice-ion coupling.

Analogous information on transition metal and rare earth ions in crystals is desirable. Seeing what can be done with the Cu^+ systems should spur the search for methods which will extract this information in these more difficult cases.

REFERENCES

1. For a review see W. B. Fowler, Physics of Color Centers, (Academic Press, New York, 1968), Ch. 2

2. F. Seitz, Rev. Mod. Phys. 23, 328 (1951).

3. C. Moore, Atomic Energy Levels, vol. 2, National Bureau of Standards Circular 557.

4. J. Simonetti and D. S. McClure, Phys. Rev. B16, 3887 (1977).
5. S. A. Payne, A. B. Goldberg and D. S. McClure, J. Chem. Phys., 78, (Part II) 3668 (1983).
6. A. B. Goldberg, D. S. McClure and C. Pedrini, Chem. Phys. Lett., 87, 508 (1982).
7. Gmelins Hanbuch der Anorg. Chem. Vol. Cu, B1, 193-195 (1958).
8. H. Bill, private communication.
9. S. A. Payne, S. Ramasesha and D. S. McClure, to be published (1984).
10. K. Fussganger, Phys. Status Solidi 34, 157 (1969), 36, 645 (1969).
11. W. Dreybrodt and K. Fussganger, ibid. 18, 133 (1966).
12. U. HOlland and F. Lüty, Phys. Rev. B19, 4298 (1979).
13. A. S. Barker and A. J. Sievers, Rev. Mod. Phys. 47 (Suppl. 2) 1-180 (1975).
14. C. Pedrini, H. Chermette, A. B. Goldberg, D. S. McClure and B. Moine, Phys. Stat. Solidi (b) 120, 753 (1983).
15. A. B. Goldberg, Thesis Princeton University (1982).
16. Ring-Ling Chien, J. Simonetti and D. S. McClure, J. Lumin. (to be published).
17. C. Pedrini (to be published).
18. C. Pedrini, private communication.
19. T. R. Bader and A. Gold, Phys. Rev., 171, 997 (1968).
20. John Berg, unpublished.
21. B. Moine and C. Pedrini, Phys. Rev. B in press (1984).
22. A. B. Goldberg, S. A. Payne and D. S. McClure, J. Chem. Phys. in press (1984).
23. S. Bell, R. Davidson and P. Warsop, J. Phys. B3, 123 (1970).
24. C. Pedrini, Phys. Stat. Sol. (B) 87, 273 (1978).
25. S. A. Payne, R. H. Austin and D. S. McClure, Phys. Rev. B29, 32 (1984).
26. S. A. Payne, A. B. Goldberg and D. S. McClure, J. Chem. Phys. in press, (1984).
27. J. G. Harrison and Chun C. Lin, Phys. Rev. B23, 3894 (1981).

Electron-Phonon Interaction and Defect Processes in Laser Materials

A.M. Stoneham

Theoretical Physics Division, Building 424.4, AERE Harwell
Didcot, Oxon OX11 ORA, Great Britain

This paper surveys the status of quantitative theoretical methods in the context of the solid state defect processes underlying tunable laser operation and performance. These include instabilities against symmetry or charge-state change, reorientation and other rate processes, including the effects of electronic excitation.

1. Introduction

The underlying mechanisms of operation of many tunable lasers, as well as their effectiveness of performance, rely on what can be described as electron-phonon coupling. This comprises a variety of solid-state processes, with the common denominator of the way in which defect states respond to changes in local atomic displacements. The present paper discusses the nature of these processes and the quantitative methods available for analysing behaviour of existing and potential laser systems. The aim of such quantitative understanding is, of course, the optimisation of laser design and performance.

The defect processes of importance can be divided into two main classes, namely those which ensure lasing occurs at all, and those which affect losses and degradation.

To be specific, consider the lasers based on colour centres in ionic alides like KCl or CaF_2 (see e.g. MOLLENAUER & OLSON 1975 [1], LIFTIN et al 1977 [2], LUTY, 1979 [3]). These are based on centres related to the F centre but with other defects or impurities adjacent to the anion vacancy and its trapped electron. Examples are a T - associated centre, the F_A(II) centre with one Li neighbour in KCl, the F_B(II) centre with two Na neighbours in KCl, the F_2^+ centre with its adjacent anion vacancy, and the F_{2A}^+ centre, for which there is an additional cation impurity as well as the anion vacancy. The effectiveness of these systems as lasers relies partly on the transition probabilities which permit population inversion and stimulated emission, and partly on the continued stability of the defects in operation. We may identify some of the key processes as these:

(i) Lattice relaxation in the excited colour centre. This may involve a symmetry change, as in the $F_A(II)$ centres, and will certainly involve the energy change leading to the Stokes shift. The energy surfaces (i.e. the total energy as a function of displacements) are also critical in deciding whether luminescence will be efficient or not (BARTRAM and STONEHAM 1975 [4]).

(ii) Energy transfer and non-radiative processes compete with luminescence. The non-radiative transitions are of three main types, namely the cooling transitions which lead to vibrational relaxation within a single electronic state; the non-radiative transitions which enable an excited electronic state to decay with the emission of heat instead of light; and the decay processes which change the nature of the colour centre itself.

(iii) Charge transfer and other decay mechanisms. Since colour centre lasers rely on specific charge states, a defect reaction such as the transfer of an electron:

$$F_A' + F_{2A} \rightarrow F_A + F_{2A}$$

is clearly going to inhibit lasing. So will processes which cause defect breakup (e.g. $F_2^+ \rightarrow F + \alpha$) or aggregation.

There have been very few detailed theoretical studies of the key processes for colour-centre laser systems. If such results were available, they could provide a route to new laser systems, e.g. those based on other halides or oxides, or to improvements in existing systems. Whilst such calculations are lacking for specific laser cases, there are detailed calculations for many key processes in closely-related systems. I shall now discuss some such cases to indicate the present range of quantitative atomistic theory. The description "quantitative" is critical, since mere qualitative explanations are often deceptive, and can be an unsound basis for future work. Likewise, the description "atomistic" is important, for continuum methods have only limited value for rough orders of magnitude.

2. Theoretical Methods

We are concerned here with three inter-related calculations. First, given a particular defect geometry, how does one calculate the wavefunction and energy (both one-electron energy and total energy)? Secondly, how does one optimise the geometry, to obtain the atomic configurations which minimise the total energy? Thirdly, how does one use this data to predict what is actually observed, i.e. thermodynamic energies, optical spectra, rate processes, etc? Many of these topics are reviewed by STONEHAM (1975) [5].

For quantitative calculations, the key lies partly in the physics and partly in the computational aspects. For present purposes I note that the most appropriate methods are real-space methods (i.e. those which recognise solids are built of atoms, as opposed to band-structure methods) used within an ionic framework. This does not mean covalency is ignored, but merely reflects a convenient choice and avoids the self-deception often encountered in relation to covalency. Ionic charges, for example, are defined in terms of dipole moment per unit displacement, and not some arbitrary division of the calculated charge density.

It is convenient to divide the methods into those for defects with no trapped carriers (or the so-called simple defects which can be regarded as such, like self-trapped holes) and those with trapped carriers. For the simple defects, without carriers, there are very efficient methods, based on the shell-model and on MOTT-LITTLETON (1938) [6] methods which embed an inner zone of ions in a dielectric continuum. These methods, which do not involve solving the Schrödinger equation, have been embodied in the general-purpose Harwell codes (HADES, PLUTO, TYPON, etc; a recent survey of applications is given in CATLOW and MACKRODT 1982 [7]) to give extremely efficient and well-tested methods. When there are trapped carriers, and when it is necessary to solve the Schrödinger equation, various options are open. Again, there are widely-used and efficient codes available, e.g. the Harwell PRISM and MOSES codes for pseudopotential methods and semi-empirical orbital methods respectively; the SERC ATMOL code is a good example of a molecular code for Hartree-Fock methods and for some degree beyond. I shall illustrate briefly the application of these various approaches to defects allied to those of colour centre lasers.

3.1 Closed-Shell Systems : KCl:Li

One of the important colour centre lasers is based on Li adjacent to an F centre in KCℓ (e.g. ref. [3]). This centre shows many interesting features, e.g. a symmetry change in the relaxed excited state. The system of isolated Li^+ substituting for K^+ in KCℓ shows a number of parallels, notably an instability such that the Li^+ moves off-centre in a (111) direction. This instability is driven by the polarisation of the $Cℓ^-$ ions by the small Li^+ ion, a factor which may contribute partly to the F_A centre symmetry change.

The results given here (mainly from SANGSTER and STONEHAM 1982 [8], with others from CATLOW et al 1978 [9]) illustrate two important points. One is the accuracy possible using empirical shell models; the other is the efficiency achievable with codes like HADES for specific properties when total energies can be obtained rapidly as a function of configuration. The empirical shell-model parameters are, of course, obtained from perfect crystal data, and there is no adjustment of any sort to enhance defect predictions.

The main results are these. First, Li should indeed go off-centre. This involves very small energies (≤ 0.1 eV). Indeed, early calculations were not especially successful (see the survey in ref. [5]), though recent systematic work on oxides (SANGSTER and STONEHAM 1981 [10]) and halides [9] has essentially complete success in predicting both whether an instability will occur and its symmetry. Secondly, the effect of hydrostatic pressure is to force Li on centre as the short-range repulsions increase. Experimentally this appears to be an abrupt transition (DEVATY and SIEVERS [11,12]) which occurs as the tunnelling frequency starts to exceed the inter-well (e.g. 111 to 111) barriers. Again, the predictions (involving energies as small as $\sim$ 0.01 eV) are in good agreement with what is seen. Thirdly, the anomalous isotope effect of the infrared spectrum gives an even more stringent test since it involves both dynamics and the nature of the phonon eigenvectors. One component observed is a peak near 40 cm^{-1} which shows an abnormal isotope effect, for the frequency rises, rather than falls, as a heavier Li isotope is substituted. This peak is predicted correctly using a supercell method. However, inspection of the eigenvectors shows the vibrations are solely of the nearby $C\ell^-$ ions, and are not infrared-active. In fact, this is the clue to the isotope-effect anomaly [8], for the 40 cm^{-1} peak can be regarded as a sideband on a tunnelling transition; since the energy is dominated by motion of ions other than Li^+, no isotope effect is expected to first order, and the small anomalous shift comes from higher-order terms.

3.2 Simple centres with one trapped electron

Here the F centre is itself the main example. The issue of accuracy shows in a different way, since most analyses concentrate on predictions of optical absorption for the range of different alkali halides. This can be obtained with reasonable success by very simple methods (e.g. point-ion or dielectric continuum models) and with very much greater success by exploiting pseudopotential methods [13]. In this widely-applied approach, ion-size corrections to the point-ion model are incorporated making use of working approximations about the closed-shell cores, namely that they do not overlap and that they are much more compact than the pseudowavefunction of the trapped electron. These assumptions, whilst really quite good for optical absorption (and indeed for allowing for energy shifts caused by nearby impurities), are not sufficient for relaxation energies, where terms are omitted comparable with those included. The answers obtained, (e.g.[14]) are broadly reasonable. Nevertheless, critical inspection suggests they are not of the desired accuracy for some applications. Even so, this pioneering approach did show how a coupled electronic structure-lattice relaxation calculation could be carried out.

More recent calculations of the relaxation in the excited state of the F centre have shown the importance of self-consistent methods, of a flexible basis, and of a consistent iteration of the lattice relaxation. Even the use of full Hartree-Fock methods has problems, some from computer capacity, some from the inherent non-linearity of the problem. One subtle point concerns the limit to which energies depend as the central region, treated in detail, is enlarged. Unless the dielectric constant is correctly predicted by the detailed atomic method, convergence to the limit will be slow, and to what may be a poor limit. Even for the F^+ centre in MgO the prediction of behaviour is non-trivial (e.g. VAIL et al 1984 [15]). This suggests strongly the use, where possible, of simple methods to get as good a starting point as is practical for fuller calculations. The technical problems of the ultimate calculations are analysed by VAIL (1984 [16]).

3.3 Components of the total energy

Assuming one could combine all the best features of the calculations, one would obtain accurate total energies for various geometries. It is these which correspond to the configuration-coordinate curves which are commonly used to interpret absorption and emission spectra, non-radiative recovery (including whether luminescence occurs or not) and the stability of a particular charge state.

Some of the critical degradation steps in laser operation involve charge transfer, e.g. of an electron to an $F_2{}^+$ or $F_{2A}{}^+$ centre (see e.g. SCHNEIDER and POLLACK 1983 [17]; GELLERMAN et al 1980 [18]). It is useful to realise the energy magnitudes involved, since many calculations make predictions only for a rigid lattice. A convenient example gives the energy needed to insert a transition metal impurity M^{N+} at a cation (Mg^{2+}) site in MgO (STONEHAM and SANGSTER 1981 [19]). There are three main terms.

One is elastic, from changes in ion size, perhaps 1-2 eV. A second is a Madelung term of about 24 Z eV (Z = N-2 here). The third is the polarisation energy, of order $6Z^2$ eV. Thus any situation involving charge state changes in an oxide may involve several cancelling energies of order 10 to 30 eV. When one notes further that, for many cases there may be three or four stable charge states in a gap of 8 eV, it is clear that no term - especially the relaxation energy - can be ignored. Indeed,that the simple pictures of one-electron levels need caution, and the question "where is the defect level in the gap" may be ill-defined. It proves advantageous to define instead the analogues of Redox potentials, E(N/N+1) such that an electron with this energy can combine with $M^{(N+1)+}$ with neither gain nor loss of energy. The accuracy with which these energies E(N/N+1) can be predicted is a major factor in determining whether particular electron-transfer reactions occur or not. Calculations for simple defects, e.g. closed-shell systems, present no problems in principle if a full and careful treatment is adopted, though simple short cuts (e.g. using cavity models with the interatomic spacing as cavity radius) can lead to errors of many eV. Similar calculations for defects with trapped carriers are in their infancy. One further word of caution is needed, namely that predictions when charge states change are less well verified than those which merely involve rearrangement of the ions. As a final point, experiment is concerned with free energies whereas much theory is concerned with thermodynamic internal energies. Since methods are available to calculate entropies, and since some of the key energies are very small (e.g. for M centre formation from F centres, ZAHRT et al 1967 [20]) it is important to recognise and estimate the correction terms in appropriate cases.

3.4 Reorientation and Diffusive processes

The simplest examples of rate processes are reorientation and diffusion. Aggregation by diffusion provides one possible route by which laser degradation can occur, and indeed is a component in the formation of F_2^+ centres (see e.g. MATSUYAMA and HIRAI 1969 [21], LUTY 1979 [3]). Some examples of predictions of diffusion rates illustrate some of the factors involved.

The electronic hole in alkali halides self-traps, producing a local distortion which effectively immobilises it. The resulting V_K centre moves by a diffusive hopping motion, as predicted for a small polaron of this type. To the extent that the V_K centre in KCℓ can be regarded as a $C\ell_2^-$ molecular species with known intra-molecular forces, the activation energy can be predicted without direct solution of the Schrödinger equation. In essence, one calculates the energy needed to distort the crystal so that the hole would have the same energy on initial or final sites. The activation energy is thus not a barrier in the usual sense. These principles, previously applied to CaF_2, $CsC\ell$ and related systems, have been examined in detail by CADE et al (1984 [22]; earlier references are cited here) for all the alkali halides. The agreement with experiment is quite good; indeed, because of the difficulties in experiment, it may be that the predictions are the more accurate.

Host crystal	Predicted activation energy (eV) for 60° jump	Observed activation energy (eV)
$KC\ell$	0.44	0.54
NaI	0.18	0.16, 0.18, 0.15
KI	0.21	0.27
RbI	0.23	0.32, 0.28, 0.31

The second example concerns self-diffusion in MgO by a vacancy mechanism. Recent calculations (SANGSTER and STONEHAM 1984 [23]) have estimated the absolute rate of diffusion, i.e. not just the activation energy, and indicate a number of complications in practice, even for closed-shell systems. These stem from the change in activation energy with thermal expansion, from the significant anharmonicity near the saddle point, and from certain ambiguities in some of the theories relating energy surfaces and vibration frequencies to rates. Nevertheless, agreement can be achieved to within experimental accuracy.

The key centres in laser systems are ones with trapped electrons. It would be optimistic to expect accurate quantitative predictions given the outstanding problems with relaxation energies. Nevertheless, we can note several features. First, in some cases, e.g. F centres, diffusion in the electronic excited state is much more rapid than in the ground state. This has parallels, e.g. for certain V_K centres (and probably for self-trapped excitons) reorientation, which is here equivalent to diffusion, occurs without barrier in an excited state. For M centres, excitation can also cause reorientation (SCHNEIDER 1970 [24]) through the $2p\pi$ to $2p\sigma$ non-radiative transition. Secondly, there is a range of processes which are enhanced in the presence of excitation. These are more important in the case of semiconductors (see e.g. STONEHAM 1981 [25]), but are clearly of concern in cases of high power densities. Thirdly, in ionic crystals, defect generation can be driven solely by optical excitation. These two related phenomena can lead to dislocation climb (see e.g. PETROFF and KIMERLING [26] for a parallel case in III-V systems) and hence an enhancement of competing non-radiative processes. Detailed modelling of the defect generation process has been carried out in one case only, namely KCℓ (ITOH et al 1977 [27]). These calculations showed considerable success, though several aspects warrant development before such approaches can be regarded as reliable methods for the wide range of possible processes of interest. Nevertheless, some general points emerged. One is that energy surfaces can be very different between ground and excited states. For the self-trapped exciton, the hole-excited states $[e\ h^*]$ proved to have much flatter energy surfaces than the electron-excited states $[e^*h]$. The shape of the surfaces - and not just the overall Stokes shift - determines the transitions among the excited states, and the degree and character (i.e. energy and polarisation) of luminescence. Moreover, transitions among excited states (e.g. $[e^*h] \rightarrow [e^*]$ prove important, and relatively modest changes in the order of the excited states can cause dramatic changes in behaviour (e.g. Itoh 1982 [28]). This system illustrates some of the main components of the important area of rapid defect processes.

4. Conclusions

The modelling of solid state processes has reached the stage at which detailed applications to solid-state laser physics would be fruitful. It is particularly clear that merely obtaining wavefunctions and one-electron energies is not enough, and that well-tested general-purpose codes are a vital factor in such applications. Theoretical modelling is an especially powerful tool in searching for new materials, for it can narrow down greatly the list of potential materials for which crystal growth techniques need to be developed. Certainly problems remain in specific areas, notably in the cases comprising coupled distortions, polarisation and self-consistent electronic structure. A systematic development of available methods and application of existing codes would give benefits of several sorts, e.g. in design of new laser systems, in the optimisation of existing systems, or in the improvement of the range of conditions (e.g. temperature) in which laser operation is possible.

References

1. L.F. Mollenauer and D.H. Olson, J. Appl. Phys. 46 3109 (1975).

2. G. Liftin, R. Beigang and H. Welling, Appl. Phys. Lett. 31 382 (1977).

3. F. Lüty, in "Optical Properties of Solids and Physics of Surfaces" (Universidad Nacional de Mexico (1979))

4. R.H. Bartram, A.M. Stoneham, Sol. St. Comm. 17 1593 (1975).

5. A.M. Stoneham, "Theory of Defects in Solids (Oxford University Press, Oxford 1975).

6. N.F. Mott and M.J. Littleton, Trans. Farad. Soc. 34 485 (1978).

7. C.R.A. Catlow and W.C. Mackrodt, "Computer Simulation of Solids", (Springer, Berlin (1982).

8. M.J.L. Sangster and A.M. Stoneham, Phys. Rev. B26 1028 (1982).

9. C.R.A. Catlow, K.M. Diller, M.J. Norgett, J. Corish, B.C.M. Parker, and P.W.M. Jacobs, Phys. Rev. B18 2739 (1978).

10. M.J.L. Sangster and A.M. Stoneham, Phil. Mag. B43 597 (1981).

11. R.P. Devaty and A.J. Sievers, Phys. Rev. B19 2343 (1979).

12. R.P. Devaty and A.J. Sievers, Phys. Rev. B22 4074 (1980).

13. R.H. Bartram, A.M. Stoneham and P.W. Gash, Phys. Rev.176 1014 (1968).

14. A.M. Stoneham and R.H. Bartram, Phys. Rev. B2 3403 (1970).

15. J.M. Vail, A.H. Harker, J.H. Harding and P. Saul, J.Phys.C17, 3401, (1984).

16. J.M. Vail, Phil. Mag., in press (1984).

17. I. Schneider and C.R. Pollock, J. Appl. Phys. 54 6193 (1983).

18. W. Gellerman, F. Lüty, K-P Koch and G. Liftin, 1980, phys. stat. sol. (9) 57 411.

19. A.M. Stoneham and M.J.L. Sangster, Phil. Mag. B43 609 (1981).

20. J.D. Zahrt, A.B. Scott and E.H. Coker, J. Chem. Phys. 46 791 (1967).

21. T. Matsuyama and M. Hirai, J. Phys. Soc. Japan 27 1526 (1969).

22. P. Cade, A.M. Stoneham and P.W. Tasker, Phys. Rev., in press (1984).

23. M.J.L. Sangster and A.M. Stoneham, J. Phys. C., to be submitted (1984).

24. I. Schneider, Phys. Rev. Lett. 24 1296 (1970).

25. A.M. Stoneham, Rep. Prog. Phys. 44 1251 (1981).

26. P.M. Petroff and L.C. Kimerling, Appl. Phys. Lett. 29 461 (1976).

27. N. Itoh, A.M. Stoneham and A.H. Harker, J. Phys. C10 4197 (1977).

28. N. Itoh, Adv. in Phys. 31 491 (1982).

Post Conference Contribution

Laser Action of H3 Color Center in Diamond

S.C. Rand and L.G. DeShazer

Hughes Research Laboratories, Malibu, CA 90265, USA

Stable room temperature laser action was achieved from the H3 color center in diamond. Laser emission peaked at 540 nm when a 2-mm-thick diamond was pumped longitudinally with 494-nm dye laser beam at 10 Hz repetition rate.

1. Introduction

An important class of tunable lasers is the color-center laser which has been developed commercially in alkali halide materials to provide widely tunable radiation in the near infrared region from 0.8 to 3.2 µm. These lasers exhibit a number of outstanding properties such as low threshold, high efficiency (∿50%) and excellent frequency and mode stability. However, they suffer disadvantages of short shelf-life, low-power optical bleaching of the laser centers, and cryogenic temperature operation. The development of a color-center laser as a practical field device has not been possible because of the need for cryogenic cooling to stabilize the color center under the effects of optical pumping. In addition, the blue-green spectral region of interest in many applications cannot be covered by F-aggregate centers for which the emission wavelengths are invariably longer than 650 nm. We report on research directed towards the goal of a color-center solid-state laser which can operate with high efficiency at room temperature.

2. Spectroscopy

We studied the absorption and fluorescence of six natural and irradiated diamonds containing different mixtures of N3 and H3 color centers. The absorption of a diamond containing both types of color centers is shown in Fig. 1 which typically gives a yellowish green color to the diamond. The H3 color center is a nitrogen substitutional defect and is extremely stable because the nitrogen atoms are an integral part of the covalent lattice. The H3 structure is two nitrogen atoms substituted for carbon, with a carbon vacancy between them as shown in Fig. 2. The H3 color centers can occur naturally or can be produced by irradiation and heat treatment of a diamond crystal. The fluorescence of the H3 center was studied and Fig. 3 shows its broad emission spectrum centered at 540 nm. We also measured the transient decay and energy transfer processes and excited-state absorption of the H3 and N3 color centers. Interaction of the H3 center with other diamond defects such as the A-aggregate center, a pair of nitrogen atoms without an adjacent vacancy, has been invoked to explain the reduction in its radiative lifetime from 18 ns to times as low as 7 ns [1]. Choice of the particular diamond sample for

* The paper reports recent work conducted after the conference. The editors felt it should be included in order to give the reader as up-to-date a picture as possible of solid state tunable laser technology.

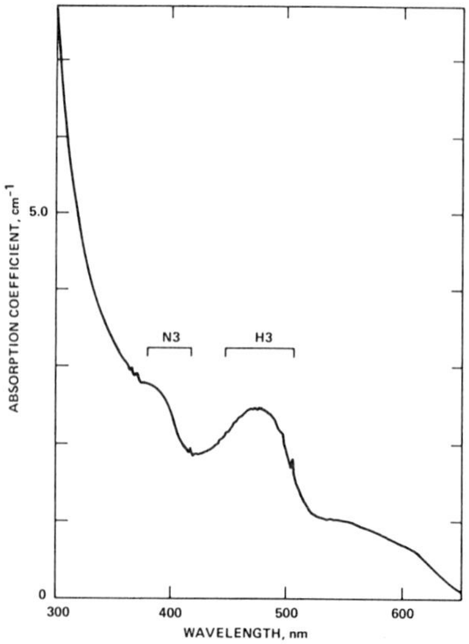

Fig. 1. Absorption spectra of H3 and N3 color centers in diamond at room temperature

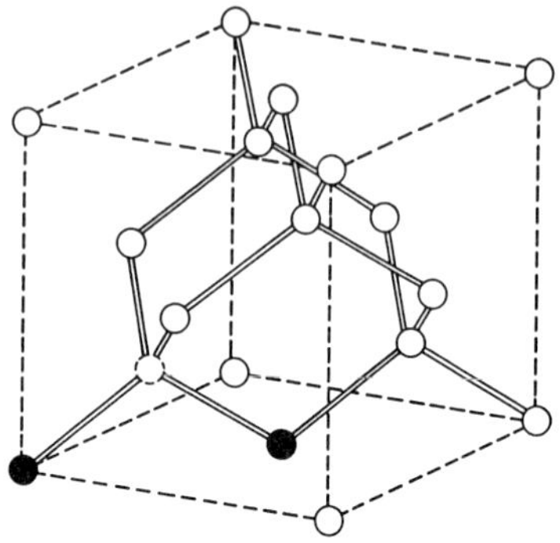

Fig. 2. Structure of H3 color center in diamond where the solid circles are nitrogen and the dashed circle is a carbon vacancy

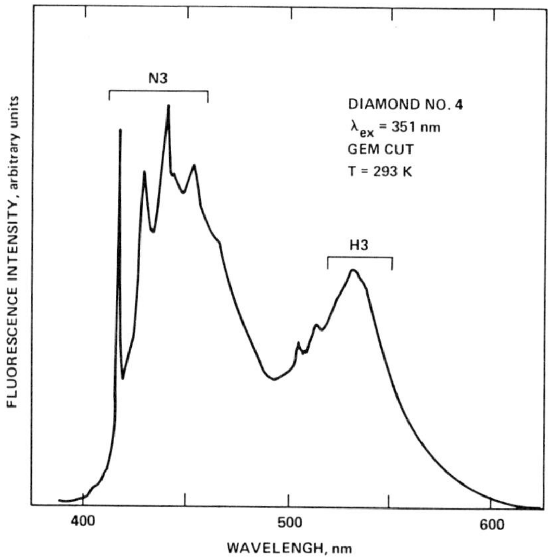

Fig. 3. Fluorescence spectra of H3 and N3 color centers in diamond at room temperature using excitation at 351 nm

laser operation depends on these considerations of energy transfer and excited state absorption. Based on these spectroscopic measurements, gain coefficients for the H3 and N3 centers were calculated to be 0.08 and 0.009 cm^{-1}, respectively.

Intense fluorescence was also observed in the 400-500 nm region from N3 centers [2], also shown in Fig. 3, but excited state absorption might be a problem for these centers. Also, energy transfer from N3 to H3 centers was observed directly, using a time resolved fluorescence technique. However, single samples containing a mixture of both N3 and H3 centers may provide an unprecedented source of optical radiation continuously tunable from 400 to 600 nm, covering almost the entire visible spectrum.

3. Laser Operation

We achieved room temperature laser action peaked at 540 nm of the H3 color center. A low-cost green diamond, 2-mm-thick and 7 mm in diameter, with polished parallel optical faces was used in the first laser experiment. The diamond laser was pumped by a dye laser beam at 494 nm where the H3 absorption coefficient was about 3 cm^{-1} in a longitudinal arrangement. The diamond was pumped by 10 Hz pulses of up to 5 mJ per pulse and laser action occurred from the 18% Fresnel reflections of the diamond faces. No bleaching of the laser-active center was observed at excitation levels as high as 70 MW/cm^2. The tuning range of this laser could ultimately be as high as 100 nm when external mirrors and tuning elements are utilized.

4. Acknowledgements

We wish to acknowledge the assistance of Dr. D.V. Manson of Gemological Institute of America, Santa Monica, California in acquiring the diamond samples and interpreting their spectroscopic properties. This research was supported by the Independent Research and Development Program of Hughes Aircraft Company.

REFERENCES

1. M.D. Crossfield et al., J. Phys. C: Solid State Phys. 7, 1909(1974)
2. M.F. Thomaz and G. Davies, Proc. Roy. Soc. Lond. A 362, 405 (1978)

Index of Contributors

Lasers

D.C. Brown

High-Peak-Power Nd: Glass Laser Systems

1981. 135 figures. IX, 276 pages. (Springer Series in Optical Sciences, Volume 25)
ISBN 3-540-10516-6

Contents: Glass Laser Physics. - Optical and Physical Properties of Laser Glasses. - Optical-Pump Sources for Nd: Glass Lasers. - Amplified Spontaneous Emission and Parasitic Oscillations in Nd: Glass Amplifiers. - Amplifiers for High-Peak-Power Nd: Glass Laser Systems. - Damage Effects in High-Peak-Power Nd: Glass Laser Systems. - Nonlinear Effects in High-Peak-Power Nd: Glass Laser Systems. - The Design of High-Peak-Power Nd: Glass Laser Systems. - Acronyms. - References. - Subject Index.

S.A. Losev

Gasdynamic Laser

1981. 100 figures. X, 297 pages. (Springer Series in Chemical Physics, Volume 12)
ISBN 3-540-10503-4

Contents: Introduction. - Basic Concepts of Quantum Electronics. - Physico-Chemical Gas Kinetics. - Relaxation in Nozzle Gas Flow. - Infrared CO_2 Gasdynamic Laser. - Gasdynamic Lasers with Other Active Medium. - Appendixes. - List of the Most Used Symbols. - References. - Subject Index.

Ecximer Lasers

Editor: **C.K. Rhodes**
2nd revised and enlarged edition. 1984. 100 figures. XII, 271 pages. (Topics in Applied Physics, Volume 30)
ISBN 3-540-13013-6

Contents: *P. W. Hoff, C.K. Rhodes:* Introduction. - *M. Krauss, F.H. Mies:* Electronic Structure and Radiative Transitions of Excimer Systems. - *M. V. McCusker:* The Rare Gas Excimers. - *C.A. Brau:* Rare Gas Halogen Excimers. - *A. Gallagher:* Metal Vapor Excimers. - *D.L. Huestis, G. Marowsky, F.K. Tittel:* Triatomic Rare-Gas-Halide Excimers. - *H. Pummer, H. Egger, C.K. Rhodes:* High-Spectral-Brightness Excimer Systems. - *K. Hohla, H. Pummer, C.K. Rhodes:* Applications of Excimer Systems. - List of Figures. - List of Tables. - Subject Index.

G. Brederlow, E. Fill, K.J. Witte

The High-Power Iodine Laser

1983. 46 figures. IX, 182 pages. (Springer Series in Optical Sciences, Volume 34)
ISBN 3-540-11792-X

Contents: Introduction. - Basic Features. - Principles of High-Power Operation. - Beam Quality and Losses. - Design and Layout of an Iodine Laser System. - The ASTERIX III System. - Scalability and Prospect of the Iodine Laser. - Conclusion. - References. - Subject Index.

Springer-Verlag
Berlin
Heidelberg
New York
Tokyo